MATH USER'S HANDBOOK

hot hot words topics

Glencoe McGraw-Hill

New York, New York
Columbus, Ohio
Chicago, Illinois
Peoria, Illinois
Woodland Hills, California

An **Online Multilingual Glossary,** which includes
11 languages in addition to English, can be found
at **www.math.glencoe.com/multilingual_glossary**

 **Glencoe**

The *McGraw-Hill* Companies

Send all inquiries to:
Glencoe/McGraw-Hill
8787 Orion Place
Columbus, OH 43240-4027

ISBN 0-07-860126-6   *Quick Review Math Handbook, Course 2*

Printed in the United States of America.
    11 12 13   027   10 09 08 07 06

HANDBOOK AT A GLANCE

iii

CONTENTS

INTRODUCTION

Why use this handbook?
You will use this mathematics handbook to refresh your
memory of concepts and skills.

What are Hot Words and how do you find them?
The Hot Words section includes a glossary of terms, a collection of
common or significant mathematical patterns, and lists of symbols
and formulas in alphabetical order. Many entries in the glossary will refer
you to chapters and topics in the Hot Topics section
for more detailed information.

4   absolute value

hot **words**

A

absolute value a number's distance
from zero on the number line
see 1•5 Integer Operations

Example:

−2 is 2 units from 0

the absolute value of −2 is 2 or |−2| = 2

accuracy the exactness of a number. For example, a number
such as 62.42812 might be rounded off to three decimal
places (62.428), to two decimal places (62.43), to one
decimal place (62.4), or to the nearest whole number
(62). The first answer is more accurate than the second,
the second more accurate than the third, and so on.
*see 2•5 Naming and Ordering Decimals, 8•1 Systems of
Measurement*

actual size the true size of an object represented by a scale
model or drawing *see 8•6 Size and Scale*

acute angle any angle that measures less than 90°
see 7•1 Naming and Classifying Angles and Triangles

Example:

∠ABC is an acute angle
0° < m∠ABC < 90°

... angle in which all angles measure less than
... Angles and Triangles

HOT WORDS

xiv

What are Hot Topics and how do you use them?

The Hot Topics section consists of nine chapters. Each chapter has several topics that give you to-the-point explanations of key mathematical concepts. Each topic includes one or more concepts. After each concept is a Check It Out section, which gives you a few problems to check your understanding of the concept. At the end of each topic, there is an exercise set.

There are problems and a vocabulary list at the beginning and end of each chapter to help you preview and review what you know.

What are Hot Solutions?

The Hot Solutions section gives you easy-to-locate answers to Check It Out and What Do You Already Know? problems.

1•2 Properties

p. 76   1. Yes   2. No   3. No   4.
p. 77   5. 26,307   6. 199   7. 0   8
9. $(3 \times 2) + (3 \times 6)$ 10. 6
p. 78   **Number Palindromes**   One
$21978 \times 4 = 87912$

1•3 Order of Operation

p. 80   1.

1•4 Fact

p. 82   1.
p. 83   3.
5.
p. 84   7
p. 85   1
p. 86
p. 87

HOT SOLUTIONS

1•4 FACTORS AND MULTIPLES

1•5

p. 90
p. 91

p. 92

1•4 Fact

Factors

Suppose that you wan rectangular pattern.

$1 \times 18 = 18$

$2 \times 9 = 18$

$3 \times 6 = 18$

Two numbers multiplied by e considered **factors** of 18. So th

To decide whether one numb there is a remainder of 0, the

FINDING THE FACT

What are the factors of 28?
• Find all pairs of numbers th product.

$1 \times 28 = 28$ $2 \times 14 = 2$

• List the factors in order, starti

The factors of 28 are 1, 2, 4, 7, 14

Check It Out

Find the factors of each nun
1. 6

2

hot words

The Hot Words section includes a glossary of terms, a collection of common or significant mathematical patterns, and lists of symbols and formulas. Many entries in the glossary will refer to chapters and topics in the Hot Topics section.

absolute value a number's distance from zero on the number line *see 1·5 Integer Operations*

Example:

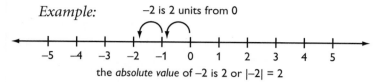

−2 is 2 units from 0

the *absolute value* of −2 is 2 or |−2| = 2

accuracy the exactness of a number. For example, a number such as 62.42812 might be rounded off to three decimal places (62.428), to two decimal places (62.43), to one decimal place (62.4), or to the nearest whole number (62). The first answer is more accurate than the second, the second more accurate than the third, and so on. *see 2·5 Naming and Ordering Decimals, 8·1 Systems of Measurement*

actual size the true size of an object represented by a scale model or drawing *see 8·6 Size and Scale*

acute angle any angle that measures less than 90° *see 7·1 Naming and Classifying Angles and Triangles*

Example:

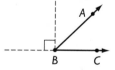

∠ABC is an *acute angle*

0° < m∠ABC < 90°

acute triangle a triangle in which all angles measure less than 90° *see 7·1 Naming and Classifying Angles and Triangles*

Example:

△ RST is an *acute triangle*

additive inverse a number that when added to a given number results in a sum of zero

Example: $(+3) + (-3) = 0$
(-3) is the *additive inverse* of 3

additive property the mathematical rule that states that if the same number is added to each side of an equation, the expressions remain equal

additive system a mathematical system in which the values of individual symbols are added together to determine the value of a sequence of symbols

Examples: The Roman numeral system, which uses symbols such as I, V, D, and M, is a well-known additive system.

This is another example of an additive system:

▽▽□

If □ equals 1 and ▽ equals 7,
then ▽▽□ equals 7 + 7 + 1 = 15

algebra a branch of mathematics in which symbols are used to represent numbers and express mathematical relationships *see Chapter 6 Algebra*

algorithm a specific step-by-step procedure for any mathematical operation *see 2•3 Addition and Subtraction of Fractions, 2•4 Multiplication and Division of Fractions, 2•6 Decimal Operations*

altitude the perpendicular distance from the base of a shape to the vertex. *Altitude* indicates the height of a shape.

Example:

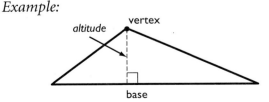

HOT WORDS

angle two rays that meet at a common endpoint
see 7•1 Naming and Classifying Angles and Triangles

Example:

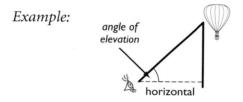

∠ ABC is formed by $\overrightarrow{BA}$ and $\overrightarrow{BC}$

angle of elevation an angle formed by an upward line of
sight and the horizontal

Example:

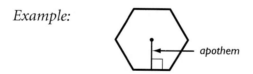

apothem a perpendicular line from the center of a regular
polygon to any one of its sides

Example:

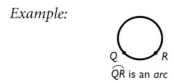

approximation an estimate of a mathematical value that is
not exact but close enough to be of use

Arabic numerals (or Hindu-Arabic numerals) the number
symbols we presently use {0, 1, 2, 3, 4, 5, 6, 7, 8, 9}

arc a section of a circle *see 7•8 Circles*

Example:

$\overset{\frown}{QR}$ is an *arc*

area the size of a surface, usually expressed in square units
*see 7•5 Area, 7•6 Surface Area, 7•8 Circles, 8•3 Area,
Volume, and Capacity*

Example:

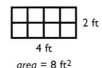

4 ft

area = 8 ft²

arithmetic expression a mathematical relationship
expressed as a number, or two or more numbers with
operation symbols
see 6•1 Writing Expressions and Equations

arithmetic sequence a mathematical progression in which
the difference between any two consecutive numbers in
the sequence is the same *see page 63*

Example: 2, 6, 10, 14, 18, 22, 26
the common difference of this *arithmetic
sequence* is 4

associative property a rule that states that the sum or
product of a set of numbers is the same, no matter how
the numbers are grouped
see 1•2 Properties, 6•2 Simplifying Expressions

Examples: $(x + y) + z = x + (y + z)$
$x \times (y \times z) = (x \times y) \times z$

average the sum of a set of values divided by the number of
values *see 4•4 Statistics*

Example: the *average* of 3, 4, 7, and 10 is
$(3 + 4 + 7 + 10) \div 4 = 6$

average speed the average rate at which an object moves

axis (pl. *axes*) [1] one of the reference lines by which a point on
a coordinate graph may be located; [2] the imaginary line
about which an object may be said to be symmetrical
(*axis* of symmetry); [3] the line about which an object
may revolve (*axis* of rotation) *see 6•7 Graphing on the
Coordinate Plane, 7•3 Symmetry and Transformations*

bar graph a way of displaying data using horizontal or vertical bars
see 4·2 Displaying Data

base [1] the side or face on which a three-dimensional shape stands; [2] the number of characters a number system contains *see 1·1 Place Value of Whole Numbers, 7·6 Surface Area, 7·7 Volume*

base-ten system the number system containing ten single-digit symbols {0, 1, 2, 3, 4, 5, 6, 7, 8, and 9} in which the numeral 10 represents the quantity ten *see 1·1 Place Value of Whole Numbers, 2·5 Naming and Ordering Decimals*

base-two system the number system containing two single-digit symbols {0 and 1} in which 10 represents the quantity two *see binary system*

benchmark a point of reference from which measurements can be made *see 2·7 Naming Percents*

best chance in a set of values, the event most likely to occur *see 4·6 Probability*

bimodal distribution a statistical model that has two different peaks of frequency distribution *see 4·3 Analyzing Data*

binary system the base two number system, in which combinations of the digits 1 and 0 represent different numbers, or values

binomial an algebraic expression that has two terms

  Examples: $x^2 + y$; $x + 1$; $a - 2b$

box plot a diagram, constructed from a set of numerical data, that shows a box indicating the middle 50% of the ranked statistics, as well as the maximum, minimum, and medium statistics *see 4·2 Displaying Data*

HOT WORDS

broken-line graph a type of line graph used to show change over a period of time *see 4·2 Displaying Data*

Example:

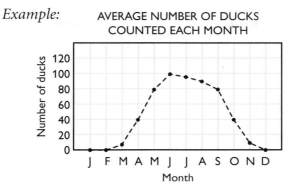

AVERAGE NUMBER OF DUCKS
COUNTED EACH MONTH

budget a spending plan based on an estimate of income and expenses *see 9·4 Spreadsheets*

cells small rectangles in a spreadsheet that hold information. Each rectangle can store a label, number, or formula *see 9·4 Spreadsheets*

center of the circle the point from which all points on a circle are equidistant *see 7·8 Circles*

chance the probability or likelihood of an occurrence, often expressed as a fraction, decimal, percentage, or ratio *see 4·6 Probability*

circle a perfectly round shape with all points equidistant from a fixed point, or center *see 7·8 Circles*

Example:

center

a *circle*

circle graph (pie chart) a way of displaying statistical data by dividing a circle into proportionally-sized "slices" *see 4•2 Displaying Data*

Example:

FAVORITE COLOR

circumference the distance around a circle, calculated by multiplying the diameter by the value pi *see 7•8 Circles*

classification the grouping of elements into separate classes or sets *see 5•3 Sets*

collinear a set of points that lie on the same line

Example:

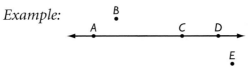

points *A, C,* and *D* are *collinear*

columns vertical lists of numbers or terms; in spreadsheets, the names of cells in a column all beginning with the same letter {A1, A2, A3, A4, . . .} *see 9•4 Spreadsheets*

combination a selection of elements from a larger set in which the order does not matter *see 4•5 Combinations and Permutations*

Example: 456, 564, and 654 are one *combination* of three digits from 4567

common denominator a whole number that is the denominator for all members of a group of fractions *see 2•3 Addition and Subtraction of Fractions*

Example: the fractions $\frac{5}{8}$ and $\frac{7}{8}$ have a *common denominator* of 8

common difference the difference between any two consecutive terms in an arithmetic sequence *see arithmetic sequence*

common factor a whole number that is a factor of each number in a set of numbers
see 1·4 Factors and Multiples

Example: 5 is a *common factor* of 10, 15, 25, and 100

common ratio the ratio of any term in a geometric sequence to the term that precedes it *see geometric sequence*

commutative property the mathematical rule that states that the order in which numbers are added or multiplied has no effect on the sum or product
see 1·2 Properties, 6·2 Simplifying Expressions

Examples: $x + y = y + x$
$x \cdot y \cdot z = y \cdot x \cdot z$

composite number a number exactly divisible by at least one whole number other than itself and 1
see 1·4 Factors and Multiples

concave polygon a polygon that has an interior angle greater than 180°

Example:

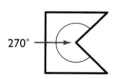

270°

a *concave polygon*

conditional a statement that something is true or will be true provided that something else is also true
see contrapositive, converse, 5·1 If/Then Statements

Example: if a polygon has three sides, then it is a triangle

cone a solid consisting of a circular base and one vertex

Example:

vertex

a *cone*

congruent figures figures that have the same size and shape. The symbol ≅ is used to indicate congruence.

Example:

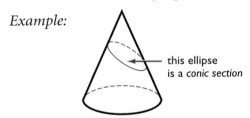

triangles *ABC* and *DEF* are *congruent*

conic section the curved shape that results when a conical surface is intersected by a plane

Example:

this ellipse
is a *conic section*

continuous data that relate to a complete range of values on the number line

Example: the possible sizes of apples are *continuous* data

contrapositive a logical equivalent of a given conditional statement, often expressed in negative terms
see 5·1 If/Then Statements

Example: "if *x*, then *y*" is a conditional statement;
"if not *y*, then not *x*" is the *contrapositive*
statement

convenience sampling a sample obtained by surveying people on the street, at a mall, or in another convenient way as opposed to a random sample *see 4·1 Collecting Data*

converse a conditional statement in which terms are expressed in reverse order *see 5·1 If/Then Statements*

Example: "if *x*, then *y*" is a conditional statement;
"if *y*, then *x*" is the *converse* statement

convex polygon a polygon that has no interior angle greater than 180°
see 7•2 Naming and Classifying Polygons and Polyhedrons

Example:

a regular hexagon is a *convex polygon*

coordinate graph the representation of points in space in relation to reference lines—usually, a horizontal *x*-axis and a vertical *y*-axis
see coordinates, 6•7 Graphing on the Coordinate Plane

coordinates an ordered pair of numbers that describes a point on a coordinate graph. The first number in the pair represents the point's distance from the origin (0, 0) along the *x*-axis, and the second represents its distance from the origin along the *y*-axis.
see ordered pairs, 6•7 Graphing on the Coordinate Plane, 6•8 Slope and Intercept

Example:

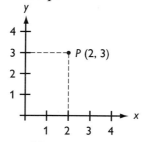

point *P* has *coordinates* (2, 3)

coplanar points or lines lying in the same plane

correlation the way in which a change in one variable corresponds to a change in another

cost an amount paid or required in payment

cost estimate an approximate amount to be paid or to be required in payment

counterexample a specific example that proves a general mathematical statement to be false
see 5·2 Counterexamples

counting numbers the set of numbers used to count objects; therefore, only those numbers that are whole and positive {1, 2, 3, 4, . . .}
see positive integers

cross product a method used to solve proportions and test whether ratios are equal: $\frac{a}{b} = \frac{c}{d}$ if $ad = bc$
see 6·5 Ratio and Proportion

cross section the figure formed by the intersection of a solid and a plane

Example:

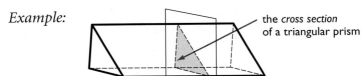

the *cross section* of a triangular prism

cube (n.) a solid figure with six square faces
see 7·2 Naming and Classifying Polygons and Polyhedrons

Example:

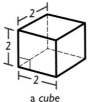

a *cube*

cube (v.) to multiply a number by itself and then by itself again *see 3·1 Powers and Exponents*

Example: $2^3 = 2 \times 2 \times 2 = 8$

cube root the number that must be multiplied by itself and then by itself again to produce a given number
see 3·2 Square and Cube Roots

Example: $\sqrt[3]{8} = 2$

cubic centimeter the amount contained in a cube with edges that are 1 cm in length *see 7·7 Volume*

cubic foot the amount contained in a cube with edges that are 1 ft in length *see 7·7 Volume*

cubic inch the amount contained in a cube with edges that are 1 in. in length *see 7·7 Volume*

cubic meter the amount contained in a cube with edges that are 1 m in length *see 7·7 Volume*

customary system units of measurement used in the United States to measure length in inches, feet, yards, and miles; capacity in cups, pints, quarts, and gallons; weight in ounces, pounds, and tons; and temperature in degrees Fahrenheit *see English system, 8·1 Systems of Measurement*

cylinder a solid shape with parallel circular bases *see 7·6 Surface Area*

Example:

a *cylinder*

decagon a plane polygon with ten angles and ten sides

decimal system the most commonly used number system, in which whole numbers and fractions are represented using base ten *see 2·5 Naming and Ordering Decimals*

Example: decimal numbers include 1,230, 1.23, 0.23, and -123

degree [1] (algebraic) the exponent of a single variable in a simple algebraic term; [2] (algebraic) the sum of the exponents of all the variables in a more complex algebraic term; [3] (algebraic) the highest degree of any term in an equation; [4] (geometric) a unit of measurement of an angle or arc, represented by the symbol °
see [1] 3•1 Powers and Exponents, [4] 7•1 Naming and Classifying Angles and Triangles, 7•8 Circles, 9•2 Scientific Calculator

> *Examples:* [1] In the term $2x^4y^3z^2$, x has a *degree* of 4, y has a *degree* of 3, and z has a *degree* of 2.
>
> [2] The term $2x^4y^3z^2$ as a whole has a *degree* of $4 + 3 + 2 = 9$.
>
> [3] The equation $x^3 = 3x^2 + x$ is an equation of the third *degree*.
>
> [4] An acute angle is an angle that measures less than 90°.

denominator the bottom number in a fraction
see 2•1 Fractions and Equivalent Fractions

> *Example:* for $\frac{a}{b}$, b is the *denominator*

dependent events a group of happenings, each of which affects the probability of the occurrence of the others
see 4•6 Probability

diagonal a line segment that connects one vertex to another (but not one next to it) on a polygon
see 7•2 Naming and Classifying Polygons and Polyhedrons

> *Example:*

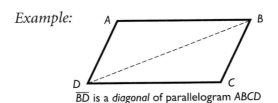

$\overline{BD}$ is a *diagonal* of parallelogram ABCD

diameter a line segment that passes through the center of a circle and divides it in half *see 7·8 Circles*

Example:

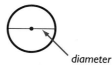

diameter

difference the result obtained when one number is subtracted from another

dimension the number of measures needed to describe a figure geometrically

Examples: A point has 0 *dimensions.*
A line or curve has 1 *dimension.*
A plane figure has 2 *dimensions.*
A solid figure has 3 *dimensions.*

direct correlation the relationship between two or more elements that increase and decrease together *see 4·3 Analyzing Data*

Example: At an hourly pay rate, an increase in the number of hours you work means an increase in the amount you get paid, while a decrease in the number of hours you work means a decrease in the amount you get paid.

discount a deduction made from the regular price of a product or service *see 2·8 Using and Finding Percents*

discrete data that can be described by whole numbers or fractional values. The opposite of *discrete* data is continuous data.

Example: the number of oranges on a tree is *discrete* data

distance the length of the shortest line segment between two points, lines, planes, and so forth *see 8·2 Length and Distance*

distance-from graph a coordinate graph that shows distance from a specified point as a function of time

distribution the frequency pattern for a set of data
see 4·3 *Analyzing Data*

distributive property of multiplication over addition
multiplication is *distributive* over addition.
For any numbers *x, y* and *z,*
$x(y + z) = xy + xz$
see 1·2 *Properties,* 6·2 *Simplifying Expressions*

double-bar graph a graphical display that uses paired
horizontal or vertical bars to show a relationship
between data see 4·2 *Displaying Data*

Example:

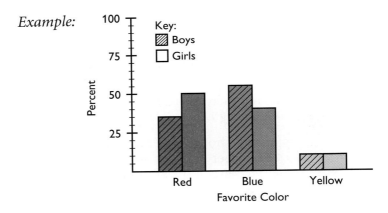

edge a line along which two planes
of a solid figure meet
see 7·2 *Naming and Classifying
Polygons and Polyhedrons*

English system units of measurement used in the
United States that measure length in inches, feet, yards,
and miles; capacity in cups, pints, quarts, and gallons;
weight in ounces, pounds, and tons; and temperature
in degrees Fahrenheit see *customary system*

equal angles angles that measure the same number of degrees
see 7·1 *Naming and Classifying Angles and Triangles*

equally likely describes outcomes or events that have the same chance of occurring *see 4•6 Probability*

equally unlikely describes outcomes or events that have the same chance of not occurring *see 4•6 Probability*

equation a mathematical sentence stating that two expressions are equal *see 6•1 Writing Expressions and Equations, 6•8 Slope and Intercept*

Example: $3 \times (7 + 8) = 9 \times 5$

equiangular having more than one angle, each of which is the same size

equiangular triangle a triangle in which each angle is 60° *see equilateral triangle, 7•1 Naming and Classifying Angles and Triangles*

equilateral a shape having more than two sides, each of which is the same length

equilateral triangle a triangle in which each side is of equal length *see equiangular triangle, 7•1 Naming and Classifying Angles and Triangles*

Example:

$AB = BC = AC$
$m\angle A = m\angle B = m\angle C = 60°$
$\triangle ABC$ is *equilateral*

equivalent equal in value

equivalent expressions expressions that always result in the same number, or have the same mathematical meaning for all replacement values of their variables *see 6•2 Simplifying Expressions*

Examples: $\frac{9}{3} + 2 = 10 - 5$

$2x + 3x = 5x$

equivalent fractions fractions that represent the same quotient but have different numerators and denominators
see 2•1 Fractions and Equivalent Fractions

Example: $\frac{5}{6} = \frac{15}{18}$

equivalent ratios ratios that are equal
see 6•5 Ratio and Proportion

Example: $\frac{5}{4} = \frac{10}{8}$; 5:4 = 10:8

estimate an approximation or rough calculation

even number any whole number that is a multiple of 2 {0, 2, 4, 6, 8, 10, 12, . . .}

event any happening to which probabilities can be assigned
see 4•6 Probability

expanded notation a method of writing a number that highlights the value of each digit
see 1•1 Place Value of Whole Numbers

Example: 867 = 800 + 60 + 7

expense an amount of money paid; cost

experimental probability a ratio that shows the total number of times the favorable outcome happened to the total number of times the experiment was done
see 4•6 Probability

exponent a numeral that indicates how many times a number or expression is to be multiplied by itself
see 1•3 Order of Operations, 3•1 Powers and Exponents, 3•3 Scientific Notation

Example: in the equation $2^3 = 8$, the *exponent* is 3

expression a mathematical combination of numbers, variables, and operations; e.g., $6x + y^2$ *see 6•1 Setting Up Expressions and Equations, 6•2 Simplifying Expressions, 6•3 Evaluating Expressions and Formulas*

*hot***words**

F

face a two-dimensional side of a
three-dimensional figure
*see 7·2 Naming and Classifying
Polygons and Polyhedrons, 7·6 Surface Area*

factor a number or expression that is multiplied by another
to yield a product *see 1·4 Factors and Multiples*

Example: 3 and 11 are *factors* of 33

factor pair two unique numbers multiplied together to yield a
product, such as 2 × 3 = 6 *see 1·4 Factors and Multiples*

factorial represented by the symbol !, the product of all the
whole numbers between 1 and a given positive whole
number *see 4·5 Combinations and Permutations*

Example: 5! = 1 × 2 × 3 × 4 × 5 = 120

fair describes a situation in which the theoretical probability
of each outcome is equal
see 4·6 Probability

Fibonacci numbers *see page 63*

flat distribution a frequency graph that shows little
difference between responses *see 4·3 Analyzing Data*

Example:

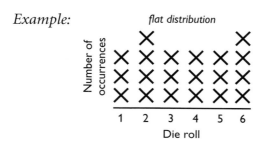

HOT WORDS

flip to "turn over" a shape
see reflection, 7·3 Symmetry and Transformations

Example:

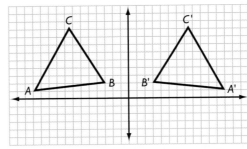

△A'B'C' is a *flip* of △ABC

forecast to predict a trend, based on statistical data
see 4·3 Analyzing Data

formula an equation that shows the relationship between two
or more quantities; a calculation performed by
spreadsheet *see pages 60–61, 6·3 Evaluating Expressions
and Formulas, 9·4 Spreadsheets*

Example: $A = \pi r^2$ is the *formula* for calculating the area
of a circle; A2 × B2 is a spreadsheet *formula*

fraction a number representing some part of a whole; a
quotient in the form $\frac{a}{b}$
see 2·1 Fractions and Equivalent Fractions

frequency graph a graph that shows similarities among the
results so one can quickly tell what is typical and what
is unusual *see 4·2 Displaying Data*

function assigns exactly one output value to each input value

Example: You are driving at 50 mi/hr. There is a
relationship between the amount of time
you drive and the distance you will travel.
You say that the distance is a *function* of
the time.

HOT WORDS

geometric sequence a sequence in which the ratio between any two consecutive terms is the same
see common ratio and page 63

> *Example:* 1, 4, 16, 64, 256, . . .
> the common ratio of this *geometric sequence* is 4

geometry the branch of mathematics concerned with the properties of figures
see Chapter 7 Geometry, 9•3 Geometry Tools

gram a metric unit used to measure mass
see 8•3 Area, Volume, and Capacity

greatest common factor (GCF) the greatest number that is a factor of two or more numbers
see 1•4 Factors and Multiples

> *Example:* 30, 60, 75
> the *greatest common factor* is 15

growth model a description of the way data change over time

harmonic sequence *see page 63*

height the distance from the base to the top of a figure
see 7•7 Volume

heptagon a polygon that has seven sides

> *Example:*

a *heptagon*

hexagon a polygon that has six sides

Example:

a *hexagon*

hexagonal prism a prism that has two hexagonal bases and six rectangular sides

Example:

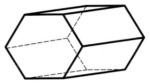

a *hexagonal prism*

hexahedron a polyhedron that has six faces

Example:

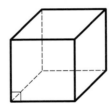

a cube is a *hexahedron*

histogram a graph in which statistical data is represented by blocks of proportionately-sized areas
see *4•2 Displaying Data*

horizontal a flat, level line or plane

hypotenuse the side of a right triangle, opposite the right angle
see *7•1 Naming and Classifying Angles and Triangles, 7•9 Pythagorean Theorem*

Example:

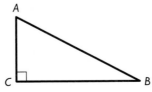

side $\overline{AB}$ is the *hypotenuse* of this right triangle

improper fraction a fraction in which the numerator is greater than the denominator *see 2·1 Fractions and Equivalent Fractions*

Examples: $\frac{21}{4}, \frac{4}{3}, \frac{2}{1}$

income the amount of money received for labor, services, or the sale of goods or property

independent event an event in which the outcome does not influence the outcome of other events *see 4·6 Probability*

inequality a statement that uses the symbols > (greater than), < (less than), ≥ (greater than or equal to), and ≤ (less than or equal to) to indicate that one quantity is larger or smaller than another *see 6·6 Inequalities*

Examples: $5 > 3$; $\frac{4}{5} < \frac{5}{4}$; $2(5 - x) > 3 + 1$

infinite, nonrepeating decimal irrational numbers, such as π and $\sqrt{2}$, that are decimals with digits that continue indefinitely but do not repeat

inscribed figure a figure that is enclosed by another figure as shown below

Examples:

a triangle *inscribed* in a circle   a circle *inscribed* in a triangle

integers the set of all whole numbers and their additive inverses {..., −5, −4, −3, −2, −1, 0, 1, 2, 3, 4, 5, ...}

intercept [1] the cutting of a line, curve, or surface by another line, curve, or surface; [2] the point at which a line or curve cuts across a given axis *see 6·8 Slope and Intercept*

HOT WORDS

intersection the set of elements that belong to each of two overlapping sets *see 5·3 Sets*

Example:

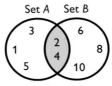

the shaded area is the *intersection* of set A (numbers 1 through 5) and set B (even numbers to 10)

inverse operations operations that undo each other

Examples: Addition and subtraction are inverse operations: $5 + 4 = 9$ and $9 - 4 = 5$. Adding 4 is the inverse of subtracting by 4. Multiplication and division are inverse operations: $5 \times 4 = 20$ and $20 \div 4 = 5$. Multiplying by 4 is the inverse of dividing by 4.

irrational numbers the set of all numbers that cannot be expressed as finite or repeating decimals

Examples: $\sqrt{2}$ (1.414214 . . .) and π (3.141592 . . .) are *irrational numbers*

isometric drawing a two-dimensional representation of a three-dimensional object in which parallel edges are drawn as parallel lines

Example:

isosceles trapezoid a trapezoid in which the two
nonparallel sides are of equal length

Example:

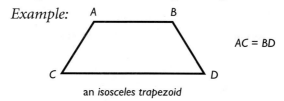

AC = BD

an *isosceles trapezoid*

isosceles triangle a triangle with at least two sides of equal
length *see 7·1 Naming and Classifying Angles and Triangles*

Example:

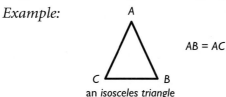

AB = AC

an *isosceles triangle*

law of large numbers when you
experiment by doing something over
and over, you get closer and closer to what
things "should" be theoretically. For example,
when you repeatedly throw a die, the proportion of 1's
that you throw will get closer to $\frac{1}{6}$ (which is the theoretical
proportion of 1's in a batch of throws).

leaf the unit-digit of an item of numerical data between 1 and 99

least common denominator (LCD) the least common
multiple of the denominators of two or more fractions
see 2·3 Addition and Subtraction of Fractions

Example: 12 is the *least common denominator* of
$\frac{1}{3}, \frac{2}{4},$ and $\frac{3}{6}$

least common multiple (LCM) the smallest nonzero whole
number that is a multiple of two or more whole numbers
*see 1·4 Factors and Multiples, 2·3 Addition and Subtraction
of Fractions*

Example: the *least common multiple* of 3, 9, and 12 is 36

HOT WORDS

legs of a triangle   the sides adjacent to the right angle of a right triangle
see 7•9 Pythagorean Theorem, 7•10 Tangent Ratio

Example:

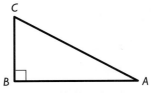

$\overline{AB}$ and $\overline{BC}$ are the *legs of a triangle ABC*

length   a measure of the distance of an object from end to end
see 8•2 Length and Distance

like terms   terms that include the same variables raised to the same powers. *Like terms* can be combined.
see 6•2 Simplifying Expressions

Example: $5x^2$ and $6x^2$ are like terms; $3xy$ and $3zy$ are not like terms

likelihood   the chance of a particular outcome occurring
see 4•6 Probability

line   a connected set of points extending forever in both directions
see 7•1 Naming and Classifying Angles and Triangles

line graph   data displayed visually to show change over time *see 4•2 Displaying Data*

Example:

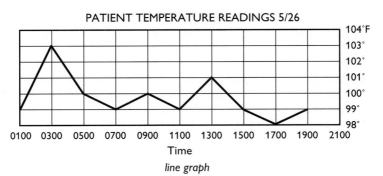

PATIENT TEMPERATURE READINGS 5/26

line graph

line of best fit on a scatter plot, a line drawn as near as possible to the various points so as to best represent the trend being graphed

Example:

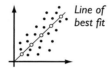

Line of best fit

line of symmetry a line along which a figure can be folded so that the two resulting halves match
see 7•3 Symmetry and Transformations

Example:

$\overline{ST}$ is a line of symmetry

line segment a section of a line running between two points
see 7•1 Naming and Classifying Angles and Triangles

Example:

A •————• B
$\overline{AB}$ is a line segment

linear equation an equation with two variables (x and y) that takes the general form $y = mx + b$, where m is the slope of the line and b is the y-intercept
see 6•4 Solving Linear Equations

linear measure the measure of the distance between two points on a line

liter a basic metric unit of capacity
see 8•3 Area, Volume, and Capacity

logic the mathematical principles that use existing theorems to prove new ones *see Chapter 5 Logic*

loss an amount of money that is lost

lowest common multiple the smallest number that is a multiple of all the numbers in a given set; same as least common multiple *see 1·4 Factors and Multiples*

   Example: for 6, 9, and 18, 18 is the *lowest common multiple*

Lucas numbers *see page 64*

magic square *see page 64*

mathematical argument a series of logical steps a person might follow to determine whether a statement is correct

maximum value the greatest value of a function or a set of numbers

mean the quotient obtained when the sum of the numbers in a set is divided by the number of addends *see average, 4·4 Statistics*

   Example: the *mean* of 3, 4, 7, and 10 is $(3 + 4 + 7 + 10) \div 4 = 6$

measurement units standard measures, such as the meter, the liter, and the gram, or the foot, the quart, and the pound *see 8·1 Systems of Measurement*

median the middle number in an ordered set of numbers *see 4·4 Statistics*

   Example: 1, 3, 9, 16, 22, 25, 27
      16 is the *median*

meter the basic metric unit of length

metric system a decimal system of weights and measurements based on the meter as its unit of length, the kilogram as its unit of mass, and the liter as its unit of capacity *see 8·1 Systems of Measurement*

midpoint the point on a line segment that divides it into two equal segments

> *Example:*
>
> A •——————•—————————• B
>   M
>
> AM = MB
> M is the midpoint of $\overline{AB}$

minimum value the least value of a function or a set of numbers

mixed number a number composed of a whole number and a fraction *see 2·3 Addition and Subtraction of Fractions*

> *Example:* $5\frac{1}{4}$

mode the number or element that occurs most frequently in a set of data *see 4·4 Statistics*

> *Example:* 1, 1, 1, 2, 2, 3, 5, 5, 6, 6, 6, 6, 8
>   6 is the *mode*

monomial an algebraic expression consisting of a single term. $5x^3y$, xy, and $2y$ are three *monomials.*

multiple the product of a given number and an integer *see 1·4 Factors and Multiples*

> *Examples:* 8 is a *multiple* of 4
>   3.6 is a *multiple* of 1.2

multiplication growth number a number that when used to multiply a given number a given number of times results in a given goal number

> *Example:* grow 10 into 40 in two steps by multiplying
>   (10 × 2 × 2 = 40)
>   2 is the *multiplication growth number*

HOT WORDS

multiplicative inverse the number for any given number that will yield 1 when the two are multiplied, same as reciprocal

Example: $10 \times \frac{1}{10} = 1$

$\frac{1}{10}$ is the multiplicative inverse of 10

natural variability the difference in results in a small number of experimental trials from the theoretical probabilities

negative integers the set of all integers that are less than zero

Examples: $-1, -2, -3, -4, -5, \ldots$

negative numbers the set of all real numbers that are less than zero

Examples: $-1, -1.36, -\sqrt{2}, -\pi$

net a two-dimensional plan that can be folded to make a three-dimensional model of a solid see 7·6 Surface Area

Example:

the net of a cube

nonagon a polygon that has nine sides

Example:

a nonagon

noncollinear not lying on the same straight line

noncoplanar not lying on the same plane

normal distribution represented by a bell curve, the most common distribution of most qualities across a given population
see 4·3 Analyzing Data

Example:

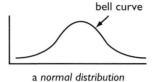

a *normal distribution*

number line a line showing numbers at regular intervals on which any real number can be indicated
see 6·6 Inequalities

Example:

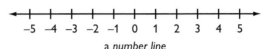

a *number line*

number symbols the symbols used in counting and measuring

Examples: $1, -\frac{1}{4}, 5, \sqrt{2}, -\pi$

number system a method of writing numbers. The Arabic *number system* is most commonly used today.

numerator the top number in a fraction. In the fraction $\frac{a}{b}$, a is the *numerator*. *see 2·1 Fractions and Equivalent Fractions*

obtuse angle any angle that measures more than 90° but less than 180° *see 7·1 Naming and Classifying Angles and Triangles*

Example:

an *obtuse angle*

HOT WORDS

obtuse triangle a triangle that has one obtuse angle
see 7·1 Naming and Classifying Angles and Triangles

Example:

△ABC is an *obtuse triangle*

octagon a polygon that has eight sides

Example:

an *octagon*

octagonal prism a prism that has two octagonal bases and eight rectangular faces

Example:

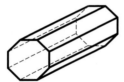

an *octagonal prism*

odd numbers the set of all integers that are not multiples of 2

odds against the ratio of the number of unfavorable outcomes to the number of favorable outcomes

odds for the ratio of the number of favorable outcomes to the number of unfavorable outcomes

one-dimensional having only one measurable quality

Example: a line and a curve are *one-dimensional*

operations arithmetical actions performed on numbers, matrices, or vectors

opposite angle in a triangle, a side and an angle are said to be opposite if the side is not used to form the angle *see 7•10 Tangent Ratio*

Example:

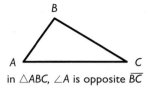

in △ABC, ∠A is opposite $\overline{BC}$

order of operations to find the answer to an equation, follow this four step process: 1) do all operations with parentheses first; 2) simplify all numbers with exponents; 3) multiply and divide in order from left to right; 4) add and subtract in order from left to right *see 1•3 Order of Operations*

ordered pair two numbers that tell the *x*-coordinate and *y*-coordinate of a point
see 6•7 Graphing on the Coordinate Plane

> *Example:* The coordinates (3, 4) are an *ordered pair*. The *x*-coordinate is 3, and the *y*-coordinate is 4.

origin the point (0, 0) on a coordinate graph where the *x*-axis and the *y*-axis intersect

orthogonal drawing always shows three views of an object—top, side, and front. The views are drawn straight-on.

Example:

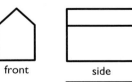

front   side

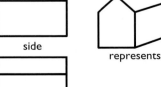

represents

top

outcome a possible result in a probability experiment

outcome grid a visual model for analyzing and representing theoretical probabilities that shows all the possible outcomes of two independent events *see 4•6 Probability*

Example:

A grid used to find the sample space for rolling a pair of dice. The outcomes are written as ordered pairs.

	1	2	3	4	5	6
1	(1, 1)	(2, 1)	(3, 1)	(4, 1)	(5, 1)	(6, 1)
2	(1, 2)	(2, 2)	(3, 2)	(4, 2)	(5, 2)	(6, 2)
3	(1, 3)	(2, 3)	(3, 3)	(4, 3)	(5, 3)	(6, 3)
4	(1, 4)	(2, 4)	(3, 4)	(4, 4)	(5, 4)	(6, 4)
5	(1, 5)	(2, 5)	(3, 5)	(4, 5)	(5, 5)	(6, 5)
6	(1, 6)	(2, 6)	(3, 6)	(4, 6)	(5, 6)	(6, 6)

There are 36 possible outcomes.

parallel straight lines or planes that remain a constant distance from each other and never intersect, represented by the symbol ∥

Example:

$\overleftrightarrow{AB}$ and $\overleftrightarrow{CD}$ are *parallel*

parallelogram a quadrilateral with two pairs of parallel sides *see 7•2 Naming and Classifying Polygons and Polyhedrons*

Example:

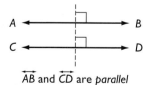

$\overline{AB} \parallel \overline{CD}$
$\overline{AC} \parallel \overline{BD}$

a *parallelogram*

parentheses the enclosing symbols (), which indicate that the terms within are a unit; for example, $(2 + 4) \div 2 = 3$

pattern a regular, repeating design or sequence of shapes or numbers *see Patterns, pages 63–65*

PEMDAS a reminder for the order of operations: 1) do all operations within **p**arentheses first; 2) simplify all numbers with **e**xponents; 3) **m**ultiply and **d**ivide in order from left to right; 4) **a**dd and **s**ubtract in order from left to right *see 1•3 Order of Operations*

pentagon a polygon that has five sides

Example:

a *pentagon*

percent a number expressed in relation to 100, represented by the symbol % *see 2•7 Meaning of Percent*

Example: 76 out of 100 students use computers
76 *percent* of students use computers

percent grade the ratio of the rise to the run of a hill, ramp, or incline expressed as a percent

Example:

percent grade = 75% ($\frac{6}{8}$)

perfect cube a number that is the cube of an integer. For example, 27 is a *perfect cube* since $27 = 3^3$.

perfect number an integer that is equal to the sum of all its positive whole number divisors, excluding the number itself

Example: $1 \times 2 \times 3 = 6$ and $1 + 2 + 3 = 6$
6 is a *perfect number*

perfect square a number that is the square of an integer. For example, 25 is a *perfect square* since $25 = 5^2$.
  see 3•2 Square and Cube Roots

perimeter the distance around the outside of a closed figure
  see 7•4 Perimeter

Example:

AB + BC + CD + DA = *perimeter*

permutation a possible arrangement of a group of objects. The number of possible arrangements of *n* objects is expressed by the term *n*!
  see factorial, 4•5 Combinations and Permutations

perpendicular two lines or planes that intersect to form a right angle

Example:

$\overline{AB}$ and $\overline{AC}$ are *perpendicular*

pi the ratio of a circle's circumference to its diameter. *Pi* is shown by the symbol π, and is approximately equal to 3.14. *see 7•8 Circles*

picture graph a graph that uses pictures or symbols to represent numbers

place value the value given to a place a digit may occupy in a numeral *see 1•1 Place Value of Whole Numbers*

place-value system a number system in which values are given to the places digits occupy in the numeral. In the decimal system, the value of each place is 10 times the value of the place to its right.
  see 1•1 Place Value of Whole Numbers

point one of four undefined terms in geometry used to define all other terms. A *point* has no size.
see 6·7 Graphing on the Coordinate Plane

polygon a simple, closed plane figure, having three or more line segments as sides
see 7·2 Naming and Classifying Polygons and Polyhedrons

Examples:

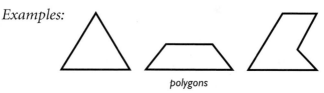

polygons

polyhedron a solid geometrical figure that has four or more plane faces
see 7·2 Naming and Classifying Polygons and Polyhedrons

Examples:

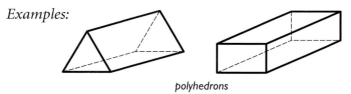

polyhedrons

population the universal set from which a sample of statistical data is selected

positive integers the set of all positive whole numbers $\{1, 2, 3, 4, 5, \ldots\}$ *see counting numbers*

positive numbers the set of all numbers that are greater than zero

Examples: $1, 1.36, \sqrt{2}, \pi$

power represented by the exponent *n*, to which a number is raised by multiplying itself *n* times
see 3·1 Powers and Exponents

Example: 7 raised to the fourth *power*
$$7^4 = 7 \times 7 \times 7 \times 7 = 2,401$$

predict to anticipate a trend by studying statistical data
see trend, 4·3 Analyzing Data

HOT WORDS

price the amount of money or goods asked for or given in exchange for something else

prime factorization the expression of a composite number as a product of its prime factors
see 1•4 Factors and Multiples

Examples: $504 = 2^3 \times 3^2 \times 7$
$30 = 2 \times 3 \times 5$

prime number a whole number greater than 1 whose only factors are 1 and itself *see 1•4 Factors and Multiples*

Examples: 2, 3, 5, 7, 11

prism a solid figure that has two parallel, congruent polygonal faces (called *bases*)
see 7•2 Naming and Classifying Polygons and Polyhedrons

Examples:

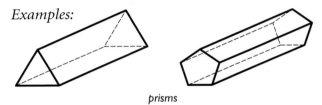

prisms

probability the study of likelihood or chance that describes the chances of an event occurring *see 4•6 Probability*

probability line a line used to order events from least likely to most likely to happen *see 4•6 Probability*

probability of events the likelihood or chance that events will occur

product the result obtained by multiplying two numbers or variables

profit the gain from a business; what is left when the cost of goods and of carrying on the business is subtracted from the amount of money taken in

project (v.) to extend a numerical model, to either greater or lesser values, in order to guess likely quantities in an unknown situation

proportion a statement that two ratios are equal
see 6•5 Ratio and Proportion

pyramid a solid geometrical figure that has a polygonal base and triangular faces that meet at a common vertex
see 7•2 Naming and Classifying Polygons and Polyhedrons

Examples:

pyramids

Pythagorean Theorem a mathematical idea stating that the sum of the squared lengths of the two shorter sides of a right triangle is equal to the squared length of the hypotenuse *see 7•9 Pythagorean Theorem*

Example:

for a right triangle, $a^2 + b^2 = c^2$

Pythagorean triple a set of three positive integers a, b, and c, such that $a^2 + b^2 = c^2$ *see 7•9 Pythagorean Theorem*

Example: for the *Pythagorean triple* $\{3, 4, 5\}$
$$3^2 + 4^2 = 5^2$$
$$9 + 16 = 25$$

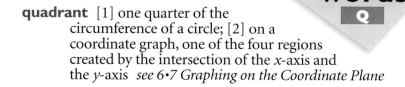

quadrant [1] one quarter of the circumference of a circle; [2] on a coordinate graph, one of the four regions created by the intersection of the x-axis and the y-axis *see 6•7 Graphing on the Coordinate Plane*

quadratic equation a polynomial equation of the second degree, generally expressed as $ax^2 + bx + c = 0$, where a, b, and c are real numbers and a is not equal to zero *see degree*

quadrilateral a polygon that has four sides
see 7·2 Naming and Classifying Polygons and Polyhedrons

Examples:

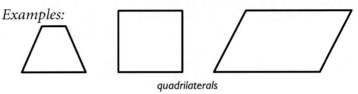

quadrilaterals

qualitative graphs a graph with words that describes such things as a general trend of profits, income, and expenses over time. It has no specific numbers.

quantitative graphs a graph that, in contrast to a qualitative graph, has specific numbers

quotient the result obtained from dividing one number or variable (the divisor) into another number or variable (the dividend)

Example:

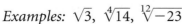

radical the indicated root of a quantity
see 3·2 Square and Cube Roots

Examples: $\sqrt{3}$, $\sqrt[4]{14}$, $\sqrt[12]{-23}$

radical sign the root symbol $\sqrt{}$

radius a line segment from the center of a circle to any point on its circumference *see 7·8 Circles*

random sample a population sample chosen so that each member has the same probability of being selected *see 4•1 Collecting Data*

range in statistics, the difference between the largest and smallest values in a sample *see 4•4 Statistics*

rank to order the data from a statistical sample on the basis of some criterion—for example, in ascending or descending numerical order *see 4•4 Statistics*

ranking the position on a list of data from a statistical sample based on some criterion

rate [1] fixed ratio between two things; [2] a comparison of two different kinds of units, for example, miles per hour or dollars per hour *see 6•5 Ratio and Proportion*

ratio a comparison of two numbers *see 6•5 Ratio and Proportion*

> *Example:* the *ratio* of consonants to vowels in the alphabet is 21:5

rational numbers the set of numbers that can be written in the form $\frac{a}{b}$, where a and b are integers and b does not equal zero

> *Examples:* $1 = \frac{1}{1}$, $\frac{2}{9}$, $3\frac{2}{7} = \frac{23}{7}$, $-.333 = -\frac{1}{3}$

ray the part of a straight line that extends infinitely in one direction from a fixed point *see 7•1 Naming and Classifying Angles and Triangles*

> *Example:*
>
>
> a *ray*

real numbers the set consisting of zero, all positive numbers, and all negative numbers. *Real numbers* include all rational and irrational numbers.

real-world data information processed by real people in everyday situations

reciprocal the result of dividing a given quantity into 1 *see 2•4 Multiplication and Division of Fractions*

> *Examples:* the *reciprocal* of 2 is $\frac{1}{2}$; of $\frac{3}{4}$ is $\frac{4}{3}$; of x is $\frac{1}{x}$

HOT WORDS

rectangle a parallelogram with four right angles
see 7·2 Naming and Classifying Polygons and Polyhedrons

Example:

a *rectangle*

rectangular prism a prism that has six rectangular faces
see 7·2 Naming and Classifying Polygons and Polyhedrons

reflection *see flip, 7·3 Symmetry and Transformations*

Example:

the *reflection* of a trapezoid

reflex angle any angle whose measure is greater than 180°
but less than 360°

Example:

A is a *reflex angle*

regular polygon a polygon in which all sides are equal and
all angles are equal

regular shape a figure in which all sides are equal and all
angles are equal

relationship a connection between two or more objects,
numbers, or sets. A mathematical *relationship* can be
expressed in words or with numbers and letters.

repeating decimal a decimal in which a digit or a set of
digits repeat infinitely

Example: 0.121212 ...

rhombus a parallelogram with all sides of equal length
    see 7·2 Naming and Classifying Polygons and Polyhedrons

      Example:

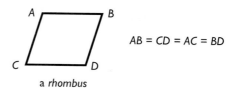

AB = CD = AC = BD

a *rhombus*

right angle an angle that measures 90° *see 7·1 Naming and Classifying Angles and Triangles*

      Example:

∠A is a *right angle*

right triangle a triangle with one right angle
    see 7·1 Naming and Classifying Angles and Triangles

      Example:

△ABC is a *right triangle*

rise the amount of vertical increase between two points
    see 6·8 Slope and Intercept

Roman numerals the numeral system consisting of the symbols I (1), V (5), X (10), L (50), C (100), D (500), and M (1,000). When a Roman symbol is preceded by a symbol of equal or greater value, the values of a symbol are added (XVI = 16). When a symbol is preceded by a symbol of lesser value, the values are subtracted (IV = 4).

root [1] the inverse of an exponent; [2] the radical sign $\sqrt{\ }$ indicates square root
    see 3·2 Square and Cube Roots

rotation a transformation in which a figure is turned a certain number of degrees around a fixed point or line *see turn, 7•3 Symmetry and Transformations*

Example:

rotation of a square

round to approximate the value of a number to a given decimal place

Examples: 2.56 rounded to the nearest tenth is 2.6;
2.54 rounded to the nearest tenth is 2.5;
365 rounded to the nearest hundred is 400

row a horizontal list of numbers or terms. In spreadsheets, the labels of cells in a *row* all end with the same number (A3, B3, C3, D3, . . .) *see 9•4 Spreadsheets*

rule a statement that describes a relationship between numbers or objects

run the horizontal distance between two points *see 6•8 Slope and Intercept*

sample a finite subset of a population, used for statistical analysis *see 4•6 Probability*

sampling with replacement a sample chosen so that each element has the chance of being selected more than once *see 4•6 Probability*

Example: A card is drawn from a deck, placed back into the deck, and a second card is drawn. Since the first card is replaced, the number of cards remains constant.

scale the ratio between the actual size of an object and a
proportional representation
see 8•6 Size and Scale

scale drawing a proportionally correct drawing of an object
or area at actual, enlarged, or reduced size
see 8•6 Size and Scale

scale factor the factor by which all the components of an
object are multiplied in order to create a proportional
enlargement or reduction
see 8•6 Size and Scale

scale size the proportional size of an enlarged or reduced
representation of an object or area *see 8•6 Size and Scale*

scalene triangle a triangle with no sides of equal length

Example:

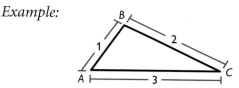

△*ABC is a scalene triangle*

scatter plot (or scatter diagram) a two-dimensional graph
in which the points corresponding to two related
factors (for example, smoking and life expectancy) are
graphed and observed for correlation
see 4•3 Analyzing Data

Example:

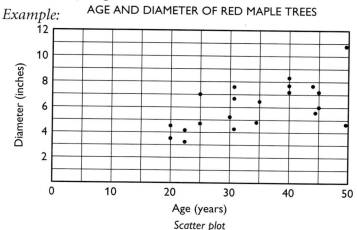

Scatter plot

scientific notation a system of writing numbers using exponents and powers of ten. A number in scientific notation is written as a number between 1 and 10 multiplied by a power of ten. *see 3•3 Scientific Notation*

> *Examples:* $9{,}572 = 9.572 \times 10^3$ and $0.00042 = 4.2 \times 10^{-4}$

segment two points and all the points on the line between them *see 7•1 Naming and Classifying Angles and Triangles*

sequence *see page 64*

series *see page 64*

side a line segment that forms an angle or joins the vertices of a polygon *see 7•4 Perimeter*

sighting measuring a length or angle of an inaccessible object by lining up a measuring tool with one's line of vision

signed number a number preceded by a positive or negative sign *see 1•5 Integer Operations*

significant digit the digit in a number that indicates its precise magnitude

> *Example:* 297,624 rounded to three significant digits is 298,000; 2.97624 rounded to three significant digits is 2.98

similar figures have the same shape but are not necessarily the same size *see 8•6 Size and Scale*

Example:

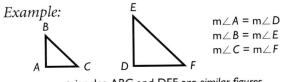

$$m\angle A = m\angle D$$
$$m\angle B = m\angle E$$
$$m\angle C = m\angle F$$

triangles ABC and DEF are *similar figures*

similarity *see similar figures*

simulation a mathematical experiment that approximates real-world process

single-bar graph a way of displaying related data using one horizontal or vertical bar to represent each data item *see 4•2 Displaying Data*

skewed distribution an asymmetrical distribution curve representing statistical data that is not balanced around the mean *see 4•3 Analyzing Data*

Example:

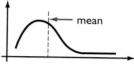

a *skewed distribution* curve

slide to move a shape to another position without rotating or reflecting it
see translation, 7•3 Symmetry and Transformations

Example:

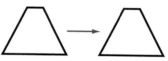

the *slide* of a trapezoid

slope [1] a way of describing the steepness of a line, ramp, hill, and so on; [2] the ratio of the rise to the run *see 6•8 Slope and Intercept*

slope angle the angle that a line forms with the x-axis or other horizontal

slope ratio the slope of a line as a ratio of the rise to the run

solid a three-dimensional shape

solution the answer to a mathematical problem. In algebra, a *solution* usually consists of a value or set of values for a variable.

special cases a number or set of numbers, such as 0, 1, fractions and negative numbers, that is considered when determining whether or not a rule is always true

speed the rate at which an object moves

HOT WORDS

speed-time graph a graph used to chart how the speed of an object changes over time

sphere a perfectly round geometric solid, consisting of a set of points equidistant from a center point

Example:

a *sphere*

spinner a device for determining outcomes in a probability experiment

Example:

a *spinner*

spiral *see page 65*

spreadsheet a computer tool where information is arranged into cells within a grid and calculations are performed within the cells. When one cell is changed, all other cells that depend on it automatically change.
see 9•4 Spreadsheets

square a rectangle with congruent sides *see 7•2 Naming and Classifying Polygons and Polyhedrons*

Example:

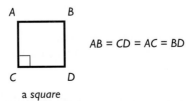

$AB = CD = AC = BD$

a *square*

square to multiply a number by itself; shown by the exponent 2
see exponent, see 3•1 Powers and Exponents

Example: $4^2 = 4 \times 4 = 16$

square centimeter a unit used to measure the size of a surface; the equivalent of a square measuring one centimeter on each side *see 8·3 Area, Volume, and Capacity*

square foot a unit used to measure the size of a surface; the equivalent of a square measuring one foot on each side *see 8·3 Area, Volume, and Capacity*

square inch a unit used to measure the size of a surface; the equivalent of a square measuring one inch on each side *see 8·3 Area, Volume, and Capacity*

square meter a unit used to measure the size of a surface; the equivalent of a square measuring one meter on each side *see 8·3 Area, Volume, and Capacity*

square number *see page 65*

Examples: 1, 4, 9, 16, 25, 36

square pyramid a pyramid with a square base

square root a number that when multiplied by itself produces a given number. For example, 3 is the *square root* of 9. *see 3·2 Square and Cube Roots*

Example: $3 \times 3 = 9$; $\sqrt{9} = 3$

square root sign the mathematical sign $\sqrt{}$; indicates that the square root of a given number is to be calculated *see 3·2 Square and Cube Roots*

standard measurement commonly used measurements, such as the meter used to measure length, the kilogram used to measure mass, and the second used to measure time *see Chapter 8 Measurement*

statistics the branch of mathematics concerning the collection and analysis of data *see 4·4 Statistics*

steepness a way of describing the amount of incline (or slope) of a ramp, hill, line, and so on

HOT WORDS

stem the ten-digit of an item of numerical data between 1 and 99 *see 4·2 Displaying Data*

stem-and-leaf plot a method of presenting numerical data between 1 and 99 by separating each number into its ten-digit (stem) and its unit-digit (leaf) and then arranging the data in ascending order of the ten-digits *see 4·2 Displaying Data*

Example:

stem	leaf
0	6
1	1 8 2 2 5
2	6 1
3	7
4	3
5	8

a *stem-and-leaf plot* for the data set 11, 26, 18, 12, 12, 15, 43, 37, 58, 6, and 21

straight angle an angle that measures 180°; a straight line

stratified random sampling a series of random samplings, each of which is taken from a specific part of the population. For example, a two-part sampling might involve taking separate samples of men and women.

strip graph a graph indicating the sequence of outcomes. A *strip graph* helps to highlight the differences among individual results and provides a strong visual representation of the concept of randomness.

Example: Outcomes of a coin toss
H = heads
T = tails

a *strip graph*

sum the result of adding two numbers or quantities

Example: $6 + 4 = 10$
10 is the *sum* of the two addends, 6 and 4

surface area the sum of the areas of all the faces of a
geometric solid, measured in square units
see 7•6 Surface Area

Example:

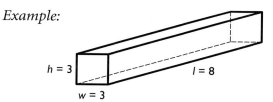

the *surface area* of this rectangular prism is
$2(3 \times 3) + 4(3 \times 8) = 114$ square units

survey a method of collecting statistical data in which people
are asked to answer questions *see 4•1 Collecting Data*

symmetry *see line of symmetry*

Example:

this hexagon has *symmetry* around the dotted line

table a collection of data arranged so
that information can be easily seen
see 4•2 Displaying Data

tally marks marks made for certain numbers of objects in
keeping account. For example, $\cancel{||||} /// = 8$.

tangent [1] a line that intersects a circle in exactly one point; [2] The *tangent* of an acute angle in a right triangle is the ratio of the length of the opposite side to the length of the adjacent side *see tangent ratio*

Example:

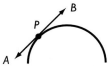

$\overleftrightarrow{AB}$ is *tangent* to the curve at point *P*

tangent ratio the ratio of the length of the side opposite a right triangle's acute angle to the length of the side adjacent to it *see 7·10 Tangent Ratio*

Example:

$$\tan S = \frac{\text{length of the side opposite to } \angle S}{\text{length of the side adjacent to } \angle S}$$

$\tan S = \frac{3}{4}$ or 0.75

tangent ratio of S is $\frac{3}{4}$ or 0.75

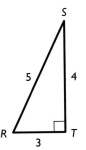

term product of numbers and variables; x, ax^2, $2x^4y^2$, and $-4ab$ are four examples of a *term*

terminating decimal a decimal with a finite number of digits

tessellation *see page 65*

Examples:

tessellations

tetrahedron a geometrical solid that has four triangular faces *see 7·2 Naming and Classifying Polygons and Polyhedrons*

Example:

a *tetrahedron*

theoretical probability the ratio of the number of favorable outcomes to the total number of possible outcomes *see 4•6 Probability*

three-dimensional having three measurable qualities: length, height, and width

tiling completely covering a plane with geometric shapes *see tessellation page 65*

time in mathematics, the element of duration, usually represented by the variable *t see 8•5 Time*

total distance the amount of space between a starting point and an endpoint, represented by *d* in the equation $d = s$ (speed) $\times t$ (time)

total distance graph a coordinate graph that shows cumulative distance traveled as a function of time

total time the duration of an event, represented by *t* in the equation $t = d$ (distance) $/ s$ (speed)

transformation a mathematical process that changes the shape or position of a geometric figure *see reflection, rotation, translation, 7•3 Symmetry and Transformations*

translation a transformation in which a geometric figure is slid to another position without rotation or reflection *see slide, 7•3 Symmetry and Transformations*

trapezoid a quadrilateral with only one pair of parallel sides *see 7•2 Naming and Classifying Polygons and Polyhedrons*

Example:

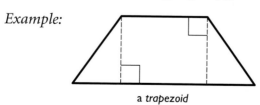

a *trapezoid*

tread the horizontal depth of one step on a stairway

tree diagram a connected, branching graph used to diagram probabilities or factors
see 1·4 Factors and Multiples, 4·5 Combinations and Permutations

Example:

a *tree diagram*

trend a consistent change over time in the statistical data representing a particular population

triangle a polygon that has three sides
see 7·1 Naming and Classifying Angles and Triangles

triangular numbers *see page 65*

triangular prism a prism with two triangular bases and three rectangular sides *see prism*

turn to move a geometric figure by rotating it around a point
see rotation, 7·3 Symmetry and Transformations

Example:

the *turning* of a triangle

two-dimensional having two measurable qualities: length and width

unequal probabilities different likelihoods of occurrence. Two events have *unequal probabilities* if one is more likely to occur than the other.

unfair where the probability of each outcome is not equal

union a set that is formed by combining the members of two or more sets, as represented by the symbol ∪. The *union* contains all members previously contained in either set *see 5·3 Sets*

Example:

Set A

3   6

9   12

Set B

4   8

12   16

Set A∪B

³ ⁴ ⁶ 8

9   16
    12

the *union* of sets A and B

unit price the price of an item expressed in a standard measure, such as *per ounce* or *per pint* or *each*

unit rate the rate in lowest terms

Example: 120 miles in two hours is equivalent to a *unit rate* of 60 miles per hour

variable a letter or other symbol that represents a number or set of numbers in an expression or an equation *see 6·1 Writing Expressions and Equations*

Example: in the equation $x + 2 = 7$, the variable is x

Venn diagram a pictorial means of representing the relationships between sets *see 5·3 Sets*

Example:

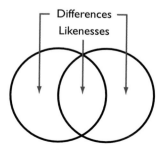

Differences

Likenesses

a *Venn diagram*

HOT WORDS

hot **words**

V

vertex (pl. *vertices*) the common point of two rays of an angle, two sides of a polygon, or three or more faces of a polyhedron

Examples:

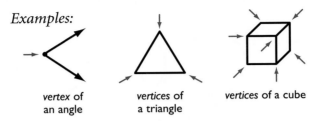

vertex of
an angle

vertices of
a triangle

vertices of a cube

vertex of tessellation the point where three or more tessellating figures come together

Example:

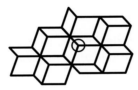

vertex of tessellation
(in the circle)

vertical a line that is perpendicular to a horizontal base line

Example:

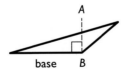

$\overline{AB}$ is *vertical* to the base
of this triangle

volume the space occupied by a solid, measured in cubic units
see 7·7 Volume

Example:

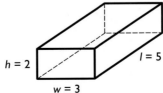

the *volume* of this rectangular prism is 30 cubic units
$2 \times 3 \times 5 = 30$

weighted average a statistical average in which each element in the sample is given a certain relative importance, or weight. For example, to find the accurate average percentage of people who own cars in three towns with different-sized populations, the largest town's percentage would have to be *weighted.*

what-if questions a question posed to frame, guide, and extend a problem

whole numbers the set of all counting numbers plus zero

Examples: 0, 1, 2, 3, 4, 5

width a measure of the distance of an object from side to side

x-axis the horizontal reference line in the coordinate graph *see 6·7 Graphing on the Coordinate Plane*

x-intercept the point at which a line or curve cuts across the *x*-axis

y-axis the vertical reference line in the coordinate graph *see 6·7 Graphing on the Coordinate Plane*

y-intercept the point at which a line or curve cuts across the *y*-axis *see 6·8 Slope and Intercept*

zero-pair one positive cube and one negative cube used to model signed number arithmetic

HOT WORDS

Formulas

Area (see 7•5)

circle $A = \pi r^2$ (pi × square of the radius)

parallelogram $A = bh$ (base × height)

rectangle $A = lw$ (length × width)

square $A = s^2$ (side squared)

trapezoid $A = \frac{1}{2}h(b_1 + b_2)$
($\frac{1}{2}$ × height × sum of the bases)

triangle $A = \frac{1}{2}bh$ ($\frac{1}{2}$ × base × height)

Volume (see 7•7)

cone $V = \frac{1}{3}\pi r^2 h$
($\frac{1}{3}$ × pi × square of the radius × height)

cylinder $V = \pi r^2 h$
(pi × square of the radius × height)

prism $V = Bh$ (area of the base × height)

pyramid $V = \frac{1}{3}Bh$ ($\frac{1}{3}$ × area of the base × height)

rectangular prism $V = lwh$ (length × width × height)

sphere $V = \frac{4}{3}\pi r^3$ ($\frac{4}{3}$ × pi × cube of the radius)

Perimeter (see 7•4)

parallelogram $P = 2a + 2b$ (2 × side a + 2 × side b)

rectangle $P = 2l + 2w$ (twice length + twice width)

square $P = 4s$ (4 × side)

triangle $P = a + b + c$ (side a + side b + side c)

Circumference (see 7•8)

circle $C = \pi d$ (pi × diameter)
or
$C = 2\pi r$ (2 × pi × radius)

Formulas

Probability (see 4•6)

The *Experimental Probability* of an event is equal to the total number of times a favorable outcome occurred, divided by the total number of times the experiment was done.

$$\frac{Experimental}{Probability} = \frac{favorable\ outcomes\ that\ occurred}{total\ number\ of\ experiments}$$

The *Theoretical Probability* of an event is equal to the number of favorable outcomes, divided by the total number of possible outcomes.

$$\frac{Theoretical}{Probability} = \frac{favorable\ outcomes}{possible\ outcomes}$$

Other

Distance	$d = rt$ (rate × time)
Interest	$i = prt$ (principle × rate × time)
PIE	Profit = Income − Expenses

FORMULAS

Symbols

{ }	set	$\overline{AB}$	segment AB
Ø	the empty set	$\overrightarrow{AB}$	ray AB
⊆	is a subset of	$\overleftrightarrow{AB}$	line AB
∪	union	$\triangle ABC$	triangle ABC
∩	intersection	$\angle ABC$	angle ABC
>	is greater than	$m\angle ABC$	measure of angle ABC
<	is less than		
≥	is greater than or equal to	AB or $m\overline{AB}$	length of segment AB
≤	is less than or equal to		
=	is equal to	$\overarc{AB}$	arc AB
≠	is not equal to	!	factorial
°	degree	$_nP_r$	permutations of n things taken r at a time
%	percent		
$f(n)$	function, f of n	$_nC_r$	combinations of n things taken r at a time
$a{:}b$	ratio of a to b, $\frac{a}{b}$		
$\lvert a \rvert$	absolute value of a	$\sqrt{}$	square root
$P(E)$	probability of an event E	$\sqrt[3]{}$	cube root
π	pi	'	foot
⊥	is perpendicular to	"	inch
‖	is parallel to	÷	divide
≅	is congruent to	/	divide
~	is similar to	*	multiply
≈	is approximately equal to	×	multiply
∠	angle	·	multiply
∟	right angle	+	add
△	triangle	−	subtract

Patterns

arithmetic sequence a sequence of numbers or terms that have a common difference between any one term and the next in the sequence. In the following sequence, the common difference is seven, so $8 - 1 = 7$; $15 - 8 = 7$; $22 - 15 = 7$, and so forth.

Example: 1, 8, 15, 22, 29, 36, 43, . . .

Fibonacci numbers a sequence in which each number is the sum of its two predecessors. Can be expressed as $x_n = x_{n-2} + x_{n-1}$. The sequence begins: 1, 1, 2, 3, 5, 8, 13, 21, 34, 55, . . .

Example:

1, 1, 2, 3, 5, 8, 13, 21, 34, 55,...
1+1=2
1+2=3
2+3=5
3+5=8

geometric sequence a sequence of terms in which each term is a constant multiple, called the *common ratio,* of the one preceding it. For instance, in nature, the reproduction of many single-celled organisms is represented by a progression of cells splitting in two in a growth progression of 1, 2, 4, 8, 16, 32, . . ., which is a geometric sequence in which the common ratio is 2.

harmonic sequence a progression $a_1, a_2, a_3, \ldots$ for which the reciprocals of the terms, $\frac{1}{a_1}, \frac{1}{a_2}, \frac{1}{a_3}, \ldots$ form an arithmetic sequence. For instance, in most musical tones, the frequencies of the sound waves are integer multiples of the fundamental frequency.

Lucas numbers a sequence in which each number is the sum of its two predecessors. Can be expressed as
$$x_n = x_{n-2} + x_{n-1}$$

The sequence begins: 1, 3, 4, 7, 11, 18, 29, 47, . . .

magic square a square array of different numbers in which rows, columns, and diagonals add up to the same total

Example:

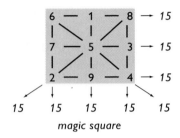

magic square

Pascal's triangle a triangular arrangement of numbers. Blaise Pascal (1623–1662) developed techniques for applying this arithmetic triangle to probability patterns.

Example:

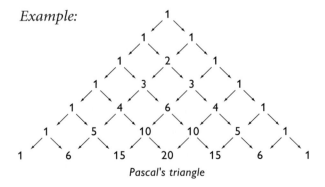

Pascal's triangle

sequence a set of elements, especially numbers, arranged in order according to some rule

series the sum of the terms of a sequence

PATTERNS

spiral a plane curve traced by a point moving around a fixed point while continuously increasing or decreasing its distance from it

Example:

the shape of a chambered nautilus shell is a *spiral*

square numbers a sequence of numbers that can be shown by dots arranged in the shape of a square. Can be expressed as x^2. The sequence begins 1, 4, 9, 16, 25, 36, 49, . . .

Example:

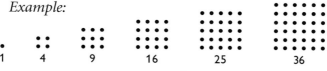

1   4   9   16   25   36

square numbers

tessellation a tiling pattern made of repeating polygons that fills a plane completely, leaving no gaps

Example:

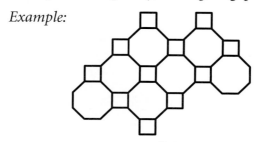

tessellation

triangular numbers a sequence of numbers that can be shown by dots arranged in the shape of a triangle. Any number in the sequence can be expressed as $x_n = x_{n-1} + n$. The sequence begins 1, 3, 6, 10, 15, 21, . . .

Example:

triangular numbers

PATTERNS

hot topics

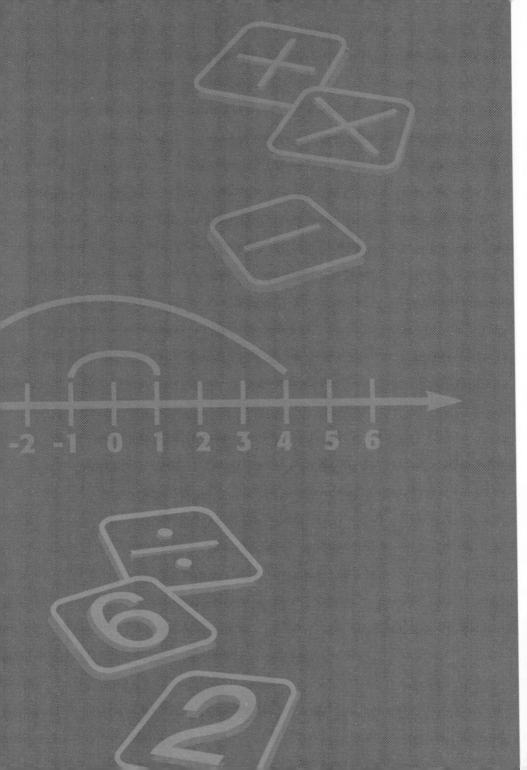

Numbers and Computation

What do you already know?

You can use the problems and the list of words that follow to see what you already know about this chapter. The answers to the problems are in Hot Solutions at the back of the book, and the definitions of the words are in Hot Words at the front of the book. You can find out more about a particular problem or word by referring to the boldfaced topic number (for example, 1•2).

Problem Set

Give the value of the 6 in each number. **1•1**
1. 267,514
2. 756,134,987

3. Write 26,748, using expanded notation. **1•1**
4. Write in order from greatest to least: 66,718; 666,718; 6,678; 596,718 **1•1**
5. Round 62,574,863 to the nearest ten, thousand, and million. **1•1**

Solve. **1•2**
6. 298×0
7. $(6 \times 3) \times 1$
8. $4,089 + 0$
9. 0×1

Solve. Use mental math if you can. **1•2**
10. $5 \times (31 + 69)$
11. $25 \times 14 \times 4$

Use parentheses to make each expression true. **1•3**
12. $4 + 7 \times 4 = 44$
13. $20 + 16 \div 4 + 5 = 29$

Is it a prime number? Write Yes or No. **1•4**
14. 57
15. 102
16. 151
17. 203

Write the prime factorization for each. **1•4**
18. 35
19. 115
20. 220

Find the GCF for each pair. **1•4**
21. 12 and 30
22. 15 and 60
23. 18 and 150

Find the LCM for each pair. **1•4**

24. 5 and 15   25. 25 and 8   26. 18 and 40

27. A mystery number is a common multiple of 2, 3, and 12. It is a factor of 108. What is the number? **1•4**

Give the absolute value of the integer. Then write its opposite. **1•5**

28. -8 29. 14
30. -11 31. 20

Add or subtract. **1•5**

32. $9 + (-2)$ 33. $6 - 7$
34. $-6 + (-6)$ 35. $4 - (-4)$
36. $-7 - (-7)$ 37. $-4 + 8$

Compute. **1•5**

38. $-5 \times (-7)$ 39. $60 \div (-12)$
40. $-48 \div (-8)$ 41. $(-4 \times 5) \times (-3)$
42. $2 \times [-8 + (-4)]$ 43. $-6 [4 - (-7)]$

44. What can you say about the product of two negative integers? **1•5**

45. What can you say about the sum of two negative integers? **1•5**

CHAPTER 1

hot **words**

absolute value   **1•5**
approximation   **1•1**
associative property   **1•2**
common factor   **1•4**
commutative property   **1•2**
composite number   **1•4**
distributive property   **1•2**
expanded notation   **1•1**
exponent   **1•4**
factor   **1•4**

greatest common factor   **1•4**
least common multiple   **1•4**
multiple   **1•4**
negative integer   **1•5**
negative number   **1•5**
number system   **1•1**
operation   **1•3**
PEMDAS   **1•3**
place value   **1•1**
positive integer   **1•5**
prime factorization   **1•4**
prime number   **1•4**
round   **1•1**

1·1 Place Value of Whole Numbers

Understanding Our Number System

You know that our **number system** is based on 10 and that the value of each place is 10 times the value of the place to its right. The value of a digit is the product of that digit and its **place value.** For instance, in the number 6,400, the 6 has a value of six thousands and the 4 has a value of four hundreds.

A *place-value chart* can help you read numbers. In a place-value chart, each group of three digits is called a *period.* Commas separate the periods. The chart below shows the diameter of Jupiter, the largest of the planets. At its equator, the distance around Jupiter is about 88,000 miles. That's about 11 times as great as Earth's diameter.

TRILLIONS PERIOD			BILLIONS PERIOD			MILLIONS PERIOD			THOUSANDS PERIOD			ONES PERIOD		
Hundred Trillions	Ten Trillions	One Trillions	Hundred Billions	Ten Billions	One Billions	Hundred Millions	Ten Millions	One Millions	Hundred Thousands	Ten Thousands	One Thousands	Hundreds	Tens	Ones
										8	8	0	0	0

To read a large number, think of the periods. At each comma, say the name of the period.

88,000 reads: eighty-eight thousand.

Check It Out

Give the value of the 4 in each number.
1. 14,083
2. 843,000,297

Write each number in words.
3. 50,326,700
4. 37,020,500,000,000

Using Expanded Notation

To show the place values of the digits in a number, you can write the number with **expanded notation.**

You can write 72,503, using expanded notation.
    72,503 = 70,000 + 2,000 + 500 + 3
- Write the ten thousands. (7 × 10,000)
- Write the thousands. (2 × 1,000)
- Write the hundreds. (5 × 100)
- Write the tens. (0 × 10)
- Write the ones. (3 × 1)

So 72,503 = (7 × 10,000) + (2 × 1,000) + (5 × 100) + (3 × 1).

Check It Out
Use expanded notation to write each number.
5. 93,045
6. 600,582

Comparing and Ordering Numbers

When you compare numbers, there are exactly three possibilities: the first number is greater than the second (3 > 2); the second is greater than the first (3 < 4); or the two numbers are equal (5 = 5). When ordering numbers, compare the numbers two at a time.

COMPARING NUMBERS

Compare 45,386 and 42,918.
- Line up the digits, starting with the ones.

    45,386

    42,918

- Beginning at the left, compare the digits in order. Find the first place where they differ.

    The digits in the thousands place differ.

- The number with the larger digit is greater.

5 > 2. So 45,386 is greater than 42,918.

Check It Out

Write >, <, or =.
7. 338,497 □ 339,006
8. 62,006 □ 61,999

Write in order from least to greatest.
9. 67,302; 62,617; 6,520; 620,009

Using Approximations

For many situations, using an **approximation** makes sense. For instance, it is reasonable to use a rounded number to express population. You might say that the population of a place is "about 40,000" rather than saying it's "39,889."

You can use this rule to **round** numbers. Look at the digit to the right of the place you are rounding to. If the digit is 5 or greater, round up. If it is less than 5, round down.

Round 518,682 to the nearest thousand.

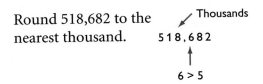

Thousands
5 1 8 , 6 8 2

6 > 5

So 518,682 rounds to 519,000.

Check It Out

10. Round 27,418 to the nearest hundred.
11. Round 528,325 to the nearest ten thousand.
12. Round 4,399,500 to the nearest million.
13. Round 574,635 to the nearest hundred thousand.

1·1 EXERCISES

Give the value of the 8 in each number.
1. 841,066
2. 624,018,257

Write each number in words.
3. 26,506,900
4. 4,030,650,000,000

Use expanded notation to write each number.
5. 49,025
6. 3,600,123

Write >, <, or =.
7. 537,257 ☐ 537,357
8. 64,106 ☐ 46,775

Write in order from least to greatest.
9. 46,388; 46,725; 14,652; 460,018

Round 58,453,533 to each place indicated.
10. nearest ten
11. nearest thousand
12. nearest hundred thousand
13. nearest ten million

Solve.
14. A hit movie took in $256,420,000. Another hit movie released around the same time took in $259,695,200. Which movie earned more money? How do you know?

15. When they rounded their sales to the nearest million, a rock group found that they sold about 3,000,000 CDs. What is the greatest number of CDs they might have sold? What is the least number?

1·2 Properties

Commutative and Associative Properties

The operations of addition and multiplication share special properties because multiplication is repeated addition.

Both addition and multiplication are **commutative.** This means that order will not change the sum or the product.

If we let a, b, and c be any whole numbers, then

$$6 + 3 = 3 + 6 \text{ and } 6 \times 3 = 3 \times 6$$
$$a + b = b + a \text{ and } a \times b = b \times a$$

Addition and multiplication can also be **associative.** This means that grouping addends or factors will not change the sum or the product.

$$(5 + 7) + 9 = 5 + (7 + 9) \text{ and } (3 \times 2) \times 4 = 3 \times (2 \times 4)$$
$$(a + b) + c = a + (b + c) \text{ and } (a \times b) \times c = a \times (b \times c)$$

Subtraction and division do not share the commutative and associative properties. For example:

$$6 - 4 = 2, \text{ but } 4 - 6 = -2; \text{ therefore } 6 - 4 \neq 4 - 6$$
$$6 \div 4 = 1.5, \text{ but } 4 \div 6 \text{ is about } 0.67;$$
therefore $6 \div 4 \neq 4 \div 6$

$$(6 - 8) - 5 = -7, \text{ but } 6 - (8 - 5) = 3;$$
therefore $(6 - 8) - 5 \neq 6 - (8 - 5)$
$$(6 \div 3) \div 4 = 0.5, \text{ but } 6 \div (3 \div 4) = 8;$$
therefore $(6 \div 3) \div 4 \neq 6 \div (3 \div 4)$

Check It Out

Write Yes or No.
1. $3 \times 4 = 4 \times 3$
2. $12 - 5 = 5 - 12$
3. $(12 \div 6) \div 3 = 12 \div (6 \div 3)$
4. $5 + (6 + 7) = (5 + 6) + 7$

Properties of One and Zero

When you add 0 to any number, the sum is that number. This is called the *zero (or identity) property of addition*. When you multiply any number by 1, the product is that number. This is called the *one (or identity) property of multiplication*. But the product of any number and 0 is 0. This is called the *zero property of multiplication*.

Check It Out

Solve.

5. $26{,}307 \times 1$
6. $199 + 0$
7. $7 \times (8 \times 0)$
8. $(4 \times 0.6) \times 1$

Distributive Property

The **distributive property** is important because it combines both addition and multiplication. This property states that multiplying a sum by a number is the same as multiplying each addend by that number and then adding the two products.

$$4(1 + 5) = (4 \times 1) + (4 \times 5)$$

If we let a, b, and c be any whole numbers, then

$$a(b + c) = ab + ac.$$

Check It Out

Rewrite each expression using the distributive property.

9. $3 \times (2 + 6)$
10. $(6 \times 7) + (6 \times 8)$

I-2 PROPERTIES

Shortcuts for Adding and Multiplying

Use the properties to help you perform some computations mentally.

$$77 + 56 + 23 = (77 + 23) + 56 = 100 + 56 = 156$$

Use commutative
and associative properties.

$$4 \times 9 \times 25 = (4 \times 25) \times 9 = 100 \times 9 = 900$$

$$8 \times 340 = (8 \times 300) + (8 \times 40) = 2,400 + 320 = 2,720$$

Use distributive property.

Number Palindromes

Do you notice anything unusual about this word, name, or sentence?

noon   Otto

Was it a can on a cat I saw?

Each one is a *palindrome*—a word, name, or sentence that reads the same forwards and backwards. It is easy to make up number palindromes using three or more digits, such as 323 or 7227. But it is harder to make up a number sentence that is the same when you read its digits from either direction, such as 10989 x 9 = 98901. Try it and see!

1·2 EXERCISES

Write Yes or No.
1. $7 \times 31 = 31 \times 7$
2. $3 \times 5 \times 6 = 3 \times 6 \times 5$
3. $4 \times 120 = (3 \times 100) \times (4 \times 20)$
4. $b \times (w + p) = bw + bp$
5. $(4 \times 6 \times 5) = (4 \times 6) + (4 \times 5)$
6. $b \times (c + d + e) = bc + bd + be$
7. $13 - 8 = 8 - 13$
8. $12 \div 4 = 4 \div 12$

Solve.
9. $42,750 \times 1$
10. $588 + 0$
11. $6 \times (0 \times 5)$
12. $0 \times 4 \times 16$
13. 1×0
14. $3.8 + 0$
15. 5.24×1
16. $(4 + 6 + 3) \times 1$

Rewrite each expression with the distributive property.
17. $5 \times (7 + 4)$
18. $(8 \times 15) + (8 \times 6)$
19. 4×550

Solve. Use mental math if you can.
20. $5 \times (24 + 6)$
21. $8 \times (22 + 78)$
22. 9×320
23. 15×8
24. $12 + 83 + 88$
25. $250 + 150 + 450$
26. 130×7
27. $11 \times 50 \times 2$
28. Give an example to show that subtraction is not associative.

29. Give an example to show that division is not commutative.

30. How would you describe the zero (or identity) property of addition?

1·3 Order of Operations

Understanding the Order of Operations

Solving a problem may involve using more than one **operation.**
Your answer will depend on the order in which you do those
operations.

For instance, take the expression $2^2 + 5 \times 6$.

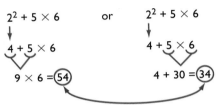

The order in which you perform
operations really makes a difference.

To make sure that there is just one answer to a series of
computations, mathematicians have agreed upon an order in
which to do the operations.

USING THE ORDER OF OPERATIONS

Simplify $4^2 - 8 \times (6 - 6)$.

- Simplify within parentheses. $= 4^2 - 8 \times (0)$
- Evaluate the exponent. $= (16) - 8 \times 0$
- Multiply or divide
  from left to right. $= 16 - (0)$
- Add or subtract from left to right. $= (16)$

So $4^2 - 8 \times (6 - 6) = 16$.

 Check It Out

Simplify.

1. $20 - 2 \times 3$ 2. $3 \times (2 + 5^2)$

1·3 EXERCISES

Is each expression true? Write Yes or No.
1. $7 \times 3 + 5 = 26$
2. $3 + 5 \times 7 = 56$
3. $6 \times (8 + 4 \div 2) = 36$
4. $6^2 - 1 = 25$
5. $(1 + 7)^2 = 64$
6. $(2^3 + 5 \times 2) + 6 = 32$
7. $45 - 5^2 = 20$
8. $(4^2 \div 4)^3 = 64$

Simplify.
9. $24 - (3 \times 5)$
10. $3 \times (4 + 5^2)$
11. $2^4 \times (10 - 7)$
12. $4^2 + (4 - 3)^2$
13. $(14 - 10)^2 \times 6$
14. $12 + 9 \times 3^2$
15. $(3^2 + 3)^2$
16. $48 \div (12 + 4)$
17. $30 - (10 - 6)^2$
18. $34 + 6 \times (4^2 \div 2)$

Use parentheses to make the expression true.
19. $5 + 5 \times 6 = 60$
20. $4 \times 25 + 75 = 400$
21. $36 \div 6 + 3 = 4$
22. $20 + 20 \div 4 - 4 = 21$
23. $10 \times 3^2 + 5 = 140$
24. $5^2 - 12 \div 3 \times 2^2 = 84$

25. Use five 2's, a set of parentheses (as needed), and any of the operations to make the numbers 1 through 3.

(**P**arentheses)
Exponents
Multiplication &
Division
Addition &
Subtraction

1•4 Factors and Multiples

Factors

Suppose that you want to arrange 18 small squares into a rectangular pattern.

$1 \times 18 = 18$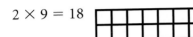

$2 \times 9 = 18$

$3 \times 6 = 18$

Two numbers multiplied by each other to produce 18 are considered **factors** of 18. So the factors of 18 are 1, 2, 3, 6, 9, and 18.

To decide whether one number is a factor of another, divide. If there is a remainder of 0, the number is a factor.

FINDING THE FACTORS OF A NUMBER

What are the factors of 28?

- Find all pairs of numbers that multiply to give the product.

  $1 \times 28 = 28 \quad 2 \times 14 = 28 \quad 4 \times 7 = 28$

- List the factors in order, starting with 1.

The factors of 28 are 1, 2, 4, 7, 14, and 28.

Check It Out

Find the factors of each number.

1. 6

2. 16

Common Factors

Factors that are the same for two or more numbers are their **common factors.**

FINDING COMMON FACTORS

What numbers are factors of both 12 and 30?
- Factors of the first number are:
  1, 2, 3, 4, 6, 12
- Factors of the second number are:
  1, 2, 3, 5, 6, 10, 15, 30
- Common factors are the numbers that are in both lists.
  1, 2, 3, 6

The common factors of 12 and 30 are 1, 2, 3, and 6.

Check It Out

List the common factors of each set of numbers.
3. 8 and 20   4. 10, 15, and 30

Greatest Common Factor

The **greatest common factor** (GCF) of two whole numbers is the greatest number that is a factor of both the numbers.

One way to find the GCF is to follow these steps:
- Find the common factors.
- Choose the greatest common factor.

What is the GCF of 16 and 40?
- Factors of 16 are 1, 2, 4, 8, 16.
- Factors of 40 are 1, 2, 4, 5, 8, 10, 20, 40.
- Common factors that are in both lists are 1, 2, 4, 8.

The greatest common factor of 16 and 40 is 8.

Check It Out

Find the GCF for each pair.
5. 8 and 20   6. 10 and 60

Divisibility Rules

There are times when you will want to know if a number is a factor of a much larger number. For instance, if you want to form teams of 3 from a group of 231 basketball players entered in a tournament, you will need to know whether 3 is a factor of 231.

> You can quickly figure out whether 231 is divisible by 3 if you know the divisibility rule for 3. A number is divisible by 3 if the sum of the digits is divisible by 3. For example, 231 is divisible by 3 because 2 + 3 + 1 = 6, and 6 is divisible by 3.

It can be helpful to know other divisibility rules. A number is divisible by:

2, if the ones digit is an even number.

3, if the sum of the digits is divisible by 3.

4, if the number formed by the last two digits is divisible by 4.

5, if the ones digit is 0 or 5.

6, if the number is divisible by 2 and 3.

8, if the number formed by the last three digits is divisible by 8.

9, if the sum of the digits is divisible by 9.

And...

Any number is divisible by **10,** if the ones digit is 0.

Check It Out

7. Is 416 divisible by 4?
8. Is 129 divisible by 9?
9. Is 462 divisible by 6?
10. Is 1,260 divisible by 5?

Prime and Composite Numbers

A **prime number** is a whole number greater than one with exactly two factors, itself and 1. Here are the first 10 prime numbers:

2, 3, 5, 7, 11, 13, 17, 19, 23, 29

Twin primes are pairs of primes whose difference is 2. (3, 5), (5, 7), and (11, 13) are examples of twin primes.

A number with more than two factors is called a **composite number.** When two composite numbers have no common factors (other than 1), they are said to be *relatively prime.* The numbers 8 and 15 are relatively prime.

One way to find out whether a number is prime or composite is to use the "sieve of Eratosthenes." Here is how it works:
- Use a chart of numbers listed in order. First skip the number 1, because it is neither prime nor composite.
- Circle the number 2 and cross out every multiple of 2.
- Next circle the number 3 and cross out every multiple of 3.
- Then continue this procedure with 5, 7, 11, and with each succeeding number that has not been crossed out.
- The prime numbers are all the circled ones. The crossed-out numbers are the composite numbers.

```
 1  (2) (3)  4  (5)  6  (7)  8   9  10
(11) 12 (13) 14 15  16 (17) 18 (19) 20
 21  22 (23) 24 25  26  27  28 (29) 30
(31) 32  33  34 35  36 (37) 38  39  40
(41) 42 (43) 44 45  46 (47) 48  49  50
 51  52 (53) 54 55  56  57  58 (59) 60...
```

 Check It Out

Is it a prime number? You can use the sieve of Eratosthenes method to decide.

11. 71   12. 87

13. 97   14. 106

Prime Factorization

Every composite number can be expressed as a product of prime factors. Use a factor tree to find the prime factors.

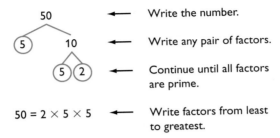

50 ←——— Write the number.

5 10 ←——— Write any pair of factors.

5 2 ←——— Continue until all factors are prime.

$50 = 2 \times 5 \times 5$ ←——— Write factors from least to greatest.

Although the order of the factors may be different because you can start with different pairs of factors, every factor tree for 50 has the same **prime factorization.** You can also use **exponents** to write the prime factorization. $50 = 2 \times 5^2$

Check It Out

What is the prime factorization of each?

15. 40   16. 100

Shortcut to Finding GCF

Use prime factorization to find the greatest common factor.

USING PRIME FACTORIZATION TO FIND THE GCF

Find the greatest common factor of 12 and 18.

- Find the prime factors of each number. Use a factor tree if it helps you.

   $12 = 2 \times 2 \times 3$ $18 = 2 \times 3 \times 3$

- Find the prime factors common to both numbers.

   2 and 3

- Find their product.

   $2 \times 3 = 6$

The GCF of 12 and 18 is 2×3, or 6.

Check It Out
Use prime factorization to find the GCF of each pair of numbers.

17. 6 and 20   18. 10 and 35
19. 15 and 42   20. 24 and 40

Multiples and Least Common Multiples

The **multiples** of a number are the whole-number products resulting when that number is a factor. In other words, you can find a multiple of a number by multiplying it by $-3, -2, -1, 0, 1, 2, 3$, and so on.

The **least common multiple** (LCM) of two numbers is the smallest positive number that is a multiple of both. One way to find the LCM of a pair of numbers is to first list multiples of each and then identify the smallest one common to both. For instance, to find the LCM of 6 and 9:
• List multiples of 6: 6, 12, 18, 24, 30, …
• List multiples of 9: 9, 18, 27, 36, …
• LCM = 18

Another way to find the LCM is to use prime factorization.

USING PRIME FACTORIZATION TO FIND THE LCM

Find the least common multiple of 6 and 9.
• Find the prime factors of each number.
$$6 = 2 \times 3 \qquad 9 = 3 \times 3$$
• Multiply the prime factors of the least number by the prime factors of the greatest number that are not factors of the least number.
$$2 \times 3 \times 3 = 18$$
The least common multiple of 6 and 9 is 18.

Check It Out
Use either method to find the LCM.

21. 6 and 8   22. 12 and 40
23. 8 and 18   24. 20 and 45

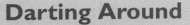

Darting Around

In the game of darts, players alternate turns, each tossing three darts at a round board. The board is divided into 20 wedges, numbered randomly from 1 to 20. You score the value of the wedge where your dart lands. The board has three rings. The bull's-eye is worth 50 points; the ring around the bull's-eye is worth 25 points; the second ring triples the value of the wedge; and the third ring doubles the value of the wedge.

Say in the first round that you throw three darts. One lands in the regular 4 space; the second lands in the second ring in the 15 wedge; and the third lands in the ring next to the bull's-eye. You earned 74 points.

Scoring works backward, by subtracting the number of points from a target number, such as 1,001 or 2,001. The record for the fewest number of darts thrown for a score of 1,001 is 19, held by Cliff Inglis (1975) and Jocky Wilson (1989). Inglis threw scores of 160, 180, 140, 180, 121, 180, and 40. Remember that Inglis's first six scores were made by tossing three darts and the last score was made by tossing one dart. What is one possible way that Inglis could have thrown his record games? See Hot Solutions for answer.

1·4 EXERCISES

Find the factors of each number.
1. 15
2. 24
3. 32
4. 56

Is it a prime number? Write Yes or No.
5. 81
6. 97
7. 107
8. 207

Write the prime factorization for each.
9. 60
10. 120
11. 160
12. 300

Find the GCF for each pair.
13. 12 and 24
14. 8 and 30
15. 18 and 45
16. 20 and 35
17. 16 and 40
18. 15 and 42

Find the LCM for each pair.
19. 6 and 7
20. 12 and 24
21. 16 and 24
22. 10 and 35

23. How do you use prime factorization to find the GCF of two numbers?
24. A mystery number is a common multiple of 2, 4, 5, and 15. It is a factor of 120. What is the number?
25. What is the divisibility rule for 8? Is 4,128 divisible by 8?

1·5 Integer Operations

Positive and Negative Integers

A glance through any newspaper shows that many quantities are expressed with **negative numbers.** For example, negative numbers show below-zero temperatures, drops in the value of stocks, or business losses.

Whole numbers greater than zero are called **positive integers.** Whole numbers less than zero are called **negative integers.**

Here is the set of all integers:

$$..., -5, -4, -3, -2, -1, 0, 1, 2, 3, 4, 5, ...$$

The integer 0 is neither positive nor negative.

Check It Out
Write an integer to describe the situation.
1. 4 below zero
2. a gain of $300

Opposites of Integers and Absolute Value

Integers can describe opposite ideas. Each integer has an opposite.

  The opposite of a gain of 4 yards is a loss of 4 yards.
  The opposite of +4 is −4.
  The opposite of spending $5 is earning $5.
  The opposite of −5 is +5.

The **absolute value** of an integer is its distance from 0 on the number line. You write the absolute value of −3 as |−3|.

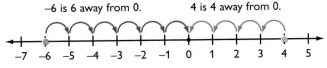

−6 is 6 away from 0.   4 is 4 away from 0.

The absolute value of 4 is 4. You write |4| = 4.
The absolute value of −6 is 6. You write |−6| = 6.

Check It Out

Give the absolute value of the integer. Then write the opposite of the original integer.

3. -15 4. $+3$

5. -12 6. 0

Adding and Subtracting Integers

Use a number line to model adding and subtracting integers.

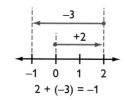

$2 + (-3) = -1$

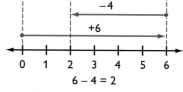

$6 - 4 = 2$

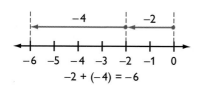

$-2 + (-4) = -6$

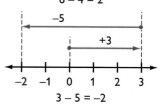

$3 - 5 = -2$

Rules for Adding or Subtracting Integers		
To	Solve	Example
Add same sign	Add absolute values. Use original sign in answer.	$-3 + (-3)$: $\|-3\| + \|-3\| = 3 + 3 = 6$ So $-3 + (-3) = -6$.
Add different signs	Subtract absolute values. Use sign of addend with greatest absolute value in answer.	$-6 + 4$: $\|-6\| - \|4\| = 6 - 4 = 2$ $\|-6\| > \|4\|$ So $-6 + 4 = -2$.
Subtract	Add opposite.	$-4 - 2 = -4 + (-2) = -6$

Check It Out

7. $5 - 8$ 8. $7 + (-7)$

9. $-8 - (-6)$ 10. $0 + (-4)$

Multiplying and Dividing Integers

Multiply and divide integers as you would with whole numbers. Then use these rules for writing the sign of the answer.

The product of two integers with like signs is positive; so is the quotient.

$$-5 \times (-3) = 15 \qquad -18 \div (-6) = 3$$

When the signs of the two integers are different, the product is negative; so is the quotient.

$$-6 \div 2 = -3 \qquad -4 \times 6 = -24 \qquad -5 \times 10 = -50$$

✔ Check It Out

Find the product or quotient.

11. $-2 \times (-4)$ 12. $12 \div (-3)$

13. $-16 \div (-4)$ 14. -7×8

Double Your Fun or Not

Did you know you can use integer math to help you in English? What is the opposite of *not inside*?

When you use two negatives in math, such as $-(-3)$, you are really asking, "What is the opposite of the number inside the parentheses?" Since the opposite of -3 is $+3$, $-(-3) = 3$. Now try the 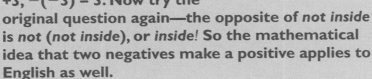 **original question again—the opposite of *not inside* is *not* (*not inside*), or *inside*! So the mathematical idea that two negatives make a positive applies to English as well.**

1·5 EXERCISES

Give the absolute value of the integer. Then write its opposite.

1. -13
2. 7
3. -5
4. 1

Add or subtract.

5. $5 - 4$
6. $4 + (-7)$
7. $-7 - (-6)$
8. $0 + (-6)$
9. $-2 + 8$
10. $0 - 9$
11. $0 - (-7)$
12. $-3 - 9$
13. $6 + (-6)$
14. $-8 - (-5)$
15. $-4 - (-4)$
16. $-8 + (-7)$

Find the product or quotient.

17. $-2 \times (-8)$
18. $9 \div (-3)$
19. $-25 \div 5$
20. -4×7
21. $5 \times (-9)$
22. $-32 \div 8$
23. $-15 \div (-3)$
24. $4 \times (-9)$

Compute.

25. $[-5 \times (-2)] \times 4$
26. $4 \times [3 \times (-4)]$
27. $[-3 \times (-4)] \times (-3)$
28. $-5 \times [3 + (-4)]$
29. $(-7 - 2) \times 3$
30. $-4 \times [6 - (-2)]$

31. Is the absolute value of a positive integer positive or negative?

32. If you know that the absolute value of an integer is 6, what are the possible values for that integer?

33. What can you say about the sum of two positive integers?

34. What can you say about the product of a negative integer and a positive integer?

35. The temperature at noon was 12°F. For the next 5 hours it dropped at a rate of 2 degrees an hour. First express this change as an integer and then give the temperature at 5 P.M.

What have you learned?

You can use the problems and the list of words that follow to see what you have learned in this chapter. You can find out more about a particular problem or word by referring to the boldfaced topic number (for example, **1•2**).

Problem Set

Give the value of the 5 in each number. **1•1**
1. 257,617
2. 785,122,907

3. Write 26,318, using expanded notation. **1•1**
4. Write in order from greatest to least: 643,254; 683,254; 6,254; 693,254 **1•1**
5. Round 48,424,492 to the nearest ten, thousand, and million. **1•1**

Solve. **1•2**

6. 836×0
7. $(5 \times 3) \times 1$
8. $6,943 + 0$
9. 0×0

Solve. Use mental math if you can. **1•2**
10. $3 \times (34 + 66)$
11. $(50 \times 12) \times 2$

Use parentheses to make each expression true. **1•3**
12. $4 + 8 \times 2 = 24$
13. $35 + 12 \div 2 + 5 = 46$

Is it a prime number? Write Yes or No. **1•4**
14. 49
15. 105
16. 163
17. 203

Write the prime factorization for each. **1•4**
18. 25
19. 170
20. 300

Find the GCF for each pair. **1•4**
21. 16 and 30
22. 12 and 50
23. 10 and 160

Find the LCM for each pair. **1•4**

24. 5 and 12

25. 15 and 8

26. 18 and 30

27. What is the divisibility rule for 5? Is 255 a multiple of 5? **1•4**

Give the absolute value of the integer. Then write its opposite. **1•5**

28. -3

29. 16

30. -12

31. 25

Add or subtract. **1•5**

32. $10 + (-8)$

33. $7 - 8$

34. $-4 + (-5)$

35. $6 - (-6)$

36. $-9 - (-9)$

37. $-6 + 14$

Compute. **1•5**

38. $-9 \times (-9)$

39. $48 \div (-12)$

40. $-27 \div (-9)$

41. $(-4 \times 3) \times (-4)$

42. $4 \times [-5 + (-7)]$

43. $-4 [5 - (-9)]$

44. What can you say about the quotient of two negative integers? **1•5**

45. What can you say about the difference of two positive integers? **1•5**

WRITE DEFINITIONS FOR THE FOLLOWING WORDS.

hot **words**

absolute value **1•5**
approximation **1•1**
associative property **1•2**
common factor **1•4**
commutative property **1•2**
composite number **1•4**
distributive property **1•2**
expanded notation **1•1**
exponent **1•4**
factor **1•4**

greatest common factor **1•4**
least common multiple **1•4**
multiple **1•4**
negative integer **1•5**
negative number **1•5**
number system **1•1**
operation **1•3**
PEMDAS **1•3**
place value **1•1**
positive integer **1•5**
prime factorization **1•4**
prime number **1•4**
round **1•1**

Fractions, Decimals, and Percents

Problem Set

1. Mr. Sebo was on a special diet. He had to carefully weigh and measure his food. He could eat 12.25 oz at breakfast, 14.621 oz at lunch, and 20.03 oz at dinner. How many ounces of food was he allowed to eat for the day? **2•6**

2. The home economics class needed $3\frac{1}{5}$ times the recipe for banana bread. How much flour is needed if the original recipe calls for $6\frac{1}{4}$ cups? **2•4**

3. Ying got 2 out of 25 problems wrong on her math quiz. What percent did she get correct? **2•8**

4. A patty made with lean ground beef is approximately 20% fat. How many ounces of fat would there be in a $14\frac{1}{2}$ oz package? **2•8**

5. Which fraction is not equivalent to $\frac{3}{12}$? **2•1**

    A. $\frac{12}{48}$ B. $\frac{1}{4}$

    C. $\frac{18}{60}$ D. $\frac{24}{96}$

Add or subtract. **2•3**

6. $\frac{4}{5} + \frac{1}{6}$ 7. $3\frac{2}{7} - 2\frac{1}{7}$

8. $5 - 1\frac{1}{2}$ 9. $2\frac{1}{5} + 3\frac{4}{7}$

10. Find the improper fraction and write it as a mixed number. **2•1**

    A. $\frac{5}{12}$ B. $\frac{3}{2}$

    C. $2\frac{1}{2}$ D. $\frac{18}{36}$

Multiply or divide. **2•4**

11. $\frac{3}{4} \times \frac{2}{5}$ 12. $\frac{2}{5} \div 2\frac{1}{3}$

13. $2\frac{1}{7} \times \frac{3}{5}$ 14. $6\frac{1}{2} \div 3\frac{1}{2}$

15. Give the place value of 5 in 34.035. **2•5**
16. Write 2.002 in expanded form. **2•5**
17. Write as a decimal: three hundred and three hundred three thousandths. **2•5**
18. Write the following numbers in order from least to greatest: 1.540; 1.504; 1.054; 0.154 **2•5**

Solve. **2•6**
19. 2.504 + 11.66
20. 10.5 − 9.06
21. 3.25 × 4.1
22. 41.76 ÷ 1.2

Use a calculator. Round to the nearest tenth. **2•8**
23. What percent of 48 is 12?
24. Find 4% of 50.
25. 23 is what percent of 25?

Write each decimal as a percent. **2•9**
26. 0.99
27. 0.4

Write each fraction as a percent. **2•9**
28. $\frac{7}{100}$
29. $\frac{83}{100}$

Write each percent as a decimal. **2•9**
30. 17%
31. 150%

Write each percent as a fraction in lowest terms. **2•9**
32. 27%
33. 120%

CHAPTER 2

hot **words**

benchmark **2•7**
common denominator **2•1**
cross product **2•1**
denominator **2•1**
discount **2•8**
equivalent **2•1**

equivalent fractions **2•1**
estimate **2•6**
factor **2•4**
fraction **2•1**
greatest common factor **2•1**
improper fraction **2•1**
least common multiple **2•1**
mixed number **2•1**

numerator **2•1**
percent **2•7**
place value **2•5**
product **2•4**
ratio **2•7**
reciprocal **2•4**
repeating decimal **2•9**
terminating decimal **2•9**
whole numbers **2•5**

WHAT DO YOU KNOW?

2·1 Fractions and Equivalent Fractions

Naming Fractions

A **fraction** can be used to name a part of a whole. For example, the flag of Bolivia is divided into three equal parts: red, yellow, and green. Each part, or color, of the Bolivian flag represents $\frac{1}{3}$ of the whole flag. $\frac{3}{3}$ or 1 represents the whole flag.

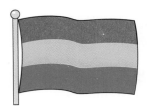

A fraction can also name part of a set.

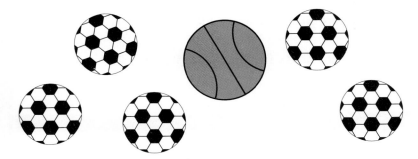

There are six balls in the set of balls. Each ball is $\frac{1}{6}$ of the set. $\frac{6}{6}$ or 1 equals the whole set. Five of the balls are soccer balls. The soccer balls represent $\frac{5}{6}$ of the set. One of the six balls is a basketball. The basketball represents $\frac{1}{6}$ of the set.

You use **numerators** and **denominators** to name fractions.

NAMING FRACTIONS

Write a fraction for the number of shaded squares.

- The denominator of the fraction tells the number of parts.

    There are 8 squares in all.
- The numerator of the fraction tells the number of parts being considered.

    5 squares are shaded.
- Write the fraction:

$$\frac{\text{parts being considered}}{\text{parts that make a whole}} = \frac{\text{numerator}}{\text{denominator}}$$

$\frac{5}{8}$ is the fraction for the number of shaded squares.

Check It Out

Write the fraction for each picture.

1. ____ of the circle is shaded.

2. ____ of the triangles are shaded.

3. Draw two pictures to represent the fraction $\frac{2}{5}$. Use regions and sets.

2·1 EQUIVALENT FRACTIONS

Methods for Finding Equivalent Fractions

Equivalent fractions are fractions that describe the same amount of a region. You can use fraction pieces to show equivalent fractions.

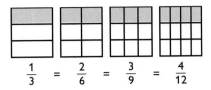

$$\frac{1}{3} = \frac{2}{6} = \frac{3}{9} = \frac{4}{12}$$

Each of the fraction pieces represents a fraction equal to $\frac{1}{3}$.

Fraction Names for One

There are an infinite number of fractions that are equal to one.

Names for one

$$\frac{4}{4} \qquad \frac{13}{13} \qquad \frac{1}{1} \qquad \frac{3{,}523}{3{,}523}$$

Any number multiplied by one is still equal to the original number. So knowing different fraction names for one can help you find equivalent fractions.

You can find a fraction that is equivalent to another fraction by multiplying the fraction by a form of one, or by dividing the numerator and the denominator by the same number.

FINDING EQUIVALENT FRACTIONS

Find a fraction equal to $\frac{4}{8}$.

Multiply by a form of one.

$$\frac{4}{8} \times \frac{3}{3} = \frac{12}{24} \qquad \frac{4}{8} = \frac{12}{24}$$

$$\frac{4 \div 2}{8 \div 2} = \frac{2}{4} \qquad \frac{4}{8} = \frac{2}{4}$$

Divide the numerator and the denominator by the same number.

Check It Out

Write two fractions equivalent to each fraction.

4. $\frac{1}{2}$ 5. $\frac{12}{24}$

6. Write three fraction names for one.

Deciding if Two Fractions Are Equivalent

Two fractions are **equivalent** if you can show that each fraction is just a different name for the same amount.

The fraction piece shows $\frac{1}{2}$ of the whole circle.

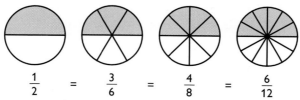

$$\frac{1}{2} \ = \ \frac{3}{6} \ = \ \frac{4}{8} \ = \ \frac{6}{12}$$

You can use fraction pieces to show names for the same amount.

One method you can use to identify equivalent fractions is to find the **cross products** of the fractions.

DECIDING IF TWO FRACTIONS ARE EQUIVALENT

Find out if $\frac{1}{2}$ and $\frac{4}{8}$ are equivalent fractions.

• Find the cross products of the fractions.

$$1 \times 8 \stackrel{?}{=} 4 \times 2$$
$$8 \stackrel{?}{=} 8$$

• Compare the cross products.

$$8 = 8$$

• If the cross products are the same, then the fractions are equivalent.

So $\frac{1}{2} = \frac{4}{8}$.

Check It Out

Use the cross-products method to determine whether or not the fractions in each pair are equivalent.

7. $\frac{3}{8}, \frac{9}{24}$ 8. $\frac{8}{12}, \frac{4}{6}$ 9. $\frac{20}{36}, \frac{5}{8}$

Least Common Denominator

Fractions that have different denominators are called *unlike fractions.* $\frac{1}{2}$ and $\frac{1}{4}$ are unlike fractions.

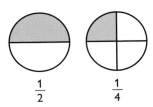

$$\frac{1}{2} \qquad \frac{1}{4}$$

Fractions can be renamed as *like fractions*. Like fractions have a **common denominator.**

$$\frac{1}{2} = \frac{2}{4} \qquad \frac{1}{4}$$

Here is one way to write fractions with common denominators.

FINDING THE LEAST COMMON DENOMINATOR

Find the least common denominator (LCD) for the fractions $\frac{2}{3}$ and $\frac{3}{5}$.

- Find the **least common multiple (LCM)** (p. 87) of the denominators.

    3: 3, 6, 9, 12, **15**, 18, 21, 24, 27

    5: 5, 10, **15**, 20, 25, 30, 35, 40, 45

    15 is the LCD for the fractions $\frac{2}{3}$ and $\frac{3}{5}$. It is the least common multiple of the denominators.

- To get equivalent fractions with the LCD, find the right form of one to multiply with each fraction.

$$\frac{2}{3} \times \frac{?}{?} = \frac{}{15} \qquad\qquad \frac{3}{5} \times \frac{?}{?} = \frac{}{15}$$

$3 \times 5 = 15$, so use $\frac{5}{5}$. $\qquad\qquad$ $5 \times 3 = 15$, so use $\frac{3}{3}$.

$$\frac{2}{3} \times \frac{5}{5} = \frac{10}{15} \qquad\qquad \frac{3}{5} \times \frac{3}{3} = \frac{9}{15}$$

Fractions $\frac{2}{3}$ and $\frac{3}{5}$ become $\frac{10}{15}$ and $\frac{9}{15}$ when renamed with their least common denominator.

Check It Out

Find the LCD. Write like fractions.

10. $\frac{3}{8}, \frac{1}{4}$ $\qquad\qquad$ 11. $\frac{7}{10}, \frac{21}{50}$

Writing Fractions in Lowest Terms

A fraction is in *lowest terms* if the *greatest common factor (GCF)* of the numerator and the denominator is 1.

These fractions are all equivalent to $\frac{3}{12}$:

$$\frac{18}{72} \qquad \frac{12}{48} \qquad \frac{6}{24} \qquad \frac{3}{12} \qquad \frac{2}{8} \qquad \frac{1}{4}$$

$$\frac{3}{12} \quad = \quad \frac{1}{4}$$

The fewest number of fraction pieces that show a fraction equivalent to $\frac{3}{12}$ is $\frac{1}{4}$. The fraction $\frac{3}{12}$ expressed in lowest terms is $\frac{1}{4}$. Another way to find lowest terms is to divide the numerator and denominator by the **greatest common factor.**

FINDING LOWEST TERM OF FRACTIONS

Express $\frac{12}{30}$ in lowest terms.

- Find the *greatest common factor* (p. 83) of the numerator and denominator.

    12: 1, 2, 3, 4, 6, 12

    30: 1, 2, 3, 5, 6, 10, 15, 30

    The GCF is 6.

- Divide the numerator and the denominator of the fraction by the GCF.

    $$\frac{12 \div 6}{30 \div 6} = \frac{2}{5}$$

$\frac{2}{5}$ is $\frac{12}{30}$ in lowest terms.

Check It Out

Express each fraction in lowest terms.

12. $\frac{2}{14}$ 13. $\frac{7}{28}$ 14. $\frac{25}{30}$

Writing Improper Fractions and Mixed Numbers

You can write fractions for amounts greater than 1. A fraction with a numerator greater than or equal to the denominator is called an **improper fraction.**

The above is an example of the improper fraction $\frac{9}{4}$. Another name for $\frac{9}{4}$ is $2\frac{1}{4}$. A whole number and a fraction make up a **mixed number,** so $2\frac{1}{4}$ is a mixed number.

You can write any mixed number as an improper fraction and any improper fraction as a mixed number.

CHANGING AN IMPROPER FRACTION TO A MIXED NUMBER

Change $\frac{15}{6}$ to a mixed number.

- Divide the numerator by the denominator.

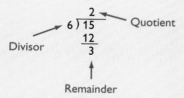

- Write the mixed number.

$$2\frac{3}{6}$$

Quotient   Remainder   Divisor

- Reduce if possible.

$$\frac{3}{6} = \frac{3 \div 3}{6 \div 3} = \frac{1}{2}$$

So $\frac{15}{6} = 2\frac{1}{2}$.

You can use multiplication to change a mixed number to an improper fraction.

CHANGING A MIXED NUMBER TO AN IMPROPER FRACTION

Change $2\frac{1}{5}$ to an improper fraction.

- Multiply the whole number part by a form of one that has the same denominator as the fraction part.

$$2 \times \frac{5}{5} = \frac{10}{5}$$

- Add the two parts.

$$2\frac{1}{5} = \frac{10}{5} + \frac{1}{5} = \frac{11}{5}$$

You can write the mixed number $2\frac{1}{5}$ as improper fracton $\frac{11}{5}$.

Check It Out

Write a mixed number for each improper fraction. Reduce if possible.

15. $\frac{19}{6}$

16. $\frac{15}{9}$

17. $\frac{16}{9}$

18. $\frac{57}{8}$

Write an improper fraction for each mixed number.

19. $8\frac{3}{4}$

20. $15\frac{1}{4}$

21. $16\frac{2}{3}$

22. $5\frac{9}{10}$

2·1 EXERCISES

Write the fraction for each picture.

1. ____ of the pieces of fruit are oranges.

2. ____ of the circle is red.

3. ____ of the triangles are green.

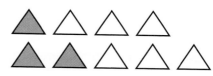

4. ____ of the balls are softballs.

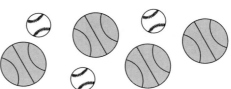

Write the fraction.

5. five eighths
6. eleven fifteenths

Write one fraction equivalent to the given fraction.

7. $\frac{3}{4}$
8. $\frac{5}{6}$
9. $\frac{10}{50}$
10. $\frac{8}{16}$
11. $\frac{12}{16}$

Express each fraction in lowest terms.

12. $\frac{28}{35}$
13. $\frac{20}{36}$
14. $\frac{66}{99}$
15. $\frac{42}{74}$
16. $\frac{22}{33}$

Write each improper fraction as a mixed number. Reduce if possible.

17. $\frac{34}{10}$
18. $\frac{38}{4}$
19. $\frac{14}{3}$
20. $\frac{44}{6}$
21. $\frac{56}{11}$

Write each mixed number as an improper fraction.

22. $12\frac{5}{6}$
23. $9\frac{7}{20}$
24. $1\frac{7}{12}$
25. $1\frac{3}{21}$

2·2 Comparing and Ordering Fractions

You can use fraction pieces to compare fractions.

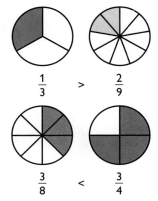

$$\frac{1}{3} \quad > \quad \frac{2}{9}$$

$$\frac{3}{8} \quad < \quad \frac{3}{4}$$

To compare fractions, you can also find *equivalent fractions* (p. 102) and compare numerators.

COMPARING FRACTIONS

Compare the fractions $\frac{5}{6}$ and $\frac{7}{8}$.

• Look at the denominators.

$$\frac{5}{⑥} \quad \text{and} \quad \frac{7}{⑧}$$

Denominators are different.

• Write equivalent fractions with a common denominator.

$$\frac{5}{6} \times \frac{4}{4} = \frac{20}{24} \qquad \frac{7}{8} \times \frac{3}{3} = \frac{21}{24}$$

• Compare the numerators.

$$20 < 21$$

• The fractions compare as the numerators compare.

$\frac{20}{24} < \frac{21}{24}$, so $\frac{5}{6} < \frac{7}{8}$.

Check It Out

Compare the fractions. Use <, >, or =.

1. $\frac{2}{9}$ □ $\frac{1}{4}$ 2. $\frac{3}{14}$ □ $\frac{2}{43}$

3. $\frac{3}{28}$ □ $\frac{4}{19}$ 4. $\frac{3}{13}$ □ $\frac{1}{6}$

Comparing Mixed Numbers

To compare mixed numbers, first compare the whole numbers. Then compare the fractions, if necessary.

COMPARING MIXED NUMBERS

Compare $1\frac{1}{5}$ and $1\frac{2}{7}$.

- Be sure the fraction parts are not improper.
  $\frac{1}{5}$ and $\frac{2}{7}$ are not improper.

- Compare the whole-number parts. If they are different, the one that is greater is the greater mixed number. If they are equal, go on.

  $1 = 1$

- To compare the fraction parts, rename them with a *common denominator* (p. 104).

  35 is the least common multiple of 5 and 7.
  Use 35 for the common denominator.

  $$\frac{1}{5} \times \frac{7}{7} = \frac{7}{35} \quad \text{and} \quad \frac{2}{7} \times \frac{5}{5} = \frac{10}{35}$$

- Compare the fractions.
  $\frac{7}{35} < \frac{10}{35}$, so $1\frac{1}{5} < 1\frac{2}{7}$.

Check It Out

Compare the mixed numbers. Use <, >, or =.

5. $3\frac{1}{2}$ □ $3\frac{5}{8}$ 6. $2\frac{3}{7}$ □ $1\frac{5}{12}$ 7. $2\frac{4}{9}$ □ $3\frac{5}{11}$

Ordering Fractions

To compare and order fractions, you can find equivalent fractions and then compare the numerators of the fractions.

ORDERING FRACTIONS WITH UNLIKE DENOMINATORS

Order the fractions $\frac{3}{5}, \frac{3}{4}$, and $\frac{7}{10}$ from least to greatest.

• Find the *least common multiple (LCM)* (p. 87) of 5, 4, and 10.

> Multiples of
> 4: 4, 8, 12, 16, (20), 24...
> 5: 5, 10, 15, (20), 25...
> 10: 10, (20), 30, 40...
> 20 is the LCM of 4, 5, and 10.

• Write equivalent fractions with a common denominator.

$$\frac{3}{5} = \frac{3}{5} \times \frac{4}{4} = \frac{12}{20}$$

$$\frac{3}{4} = \frac{3}{4} \times \frac{5}{5} = \frac{15}{20}$$

$$\frac{7}{10} = \frac{7}{10} \times \frac{2}{2} = \frac{14}{20}$$

• The fractions compare as the numerators compare.

$$\frac{12}{20} < \frac{14}{20} < \frac{15}{20}, \text{ so } \frac{3}{5} < \frac{7}{10} < \frac{3}{4}.$$

Check It Out

Order the fractions from least to greatest.

8. $\frac{5}{6}; \frac{5}{7}; \frac{3}{4}; \frac{2}{3}$

9. $\frac{2}{3}; \frac{9}{10}; \frac{7}{8}; \frac{3}{4}$

10. $\frac{1}{8}; \frac{3}{4}; \frac{5}{12}; \frac{3}{8}; \frac{5}{6}$

2·2 EXERCISES

Compare each fraction. Use $<$, $>$, or $=$.

1. $\frac{7}{12}$ $\square$ $\frac{5}{6}$
2. $\frac{2}{3}$ $\square$ $\frac{5}{9}$
3. $\frac{3}{8}$ $\square$ $\frac{1}{3}$
4. $\frac{1}{6}$ $\square$ $\frac{2}{9}$
5. $\frac{8}{9}$ $\square$ $\frac{17}{18}$

Compare each mixed number. Use $<$, $>$, or $=$.

6. $3\frac{2}{5}$ $\square$ $2\frac{4}{5}$
7. $1\frac{2}{3}$ $\square$ $1\frac{5}{9}$
8. $2\frac{4}{7}$ $\square$ $2\frac{5}{12}$
9. $4\frac{2}{5}$ $\square$ $4\frac{3}{7}$

Order the fractions and mixed numbers from least to greatest.

10. $\frac{3}{8}, \frac{2}{5}, \frac{7}{20}$
11. $\frac{2}{6}, \frac{8}{21}, \frac{4}{14}$
12. $\frac{7}{12}, \frac{23}{40}, \frac{8}{15}, \frac{19}{30}$
13. $1\frac{8}{11}, 2\frac{1}{4}, 1\frac{3}{4}$

Use the following information to answer items 14 and 15.

RECESS SHOTS THAT MADE BASKETS

Toshi $\frac{5}{7}$

Vanessa $\frac{8}{12}$

Sylvia $\frac{4}{9}$

Derrick $\frac{7}{10}$

14. Who was more accurate in shots, Vanessa or Toshi?
15. Order the players from least accurate to most accurate shots.

2·3 Addition and Subtraction of Fractions

Adding and Subtracting Fractions with Like Denominators

When you add or subtract fractions that have the same, or like, denominators, you add or subtract only the numerators. The denominators stay the same.

$$\frac{1}{9} + \frac{2}{9} = \frac{3}{9}$$

You can use fraction drawings to model the addition and subtraction of fractions with like denominators.

ADDING AND SUBTRACTING FRACTIONS WITH LIKE DENOMINATORS

Add $\frac{5}{12} + \frac{11}{12}$.

- Add or subtract the numerators.

$$5 + 11 = 16$$

- Write the result over the denominator.

$$\frac{5}{12} + \frac{11}{12} = \frac{16}{12}$$

- Simplify, if possible.

$$\frac{16}{12} = 1\frac{4}{12} = 1\frac{1}{3}$$

$$\frac{5}{12} + \frac{11}{12} = 1\frac{1}{3}$$

Check It Out

Add or subtract. Simplify, if possible.

1. $\frac{4}{5} + \frac{3}{5}$

2. $\frac{11}{12} + \frac{7}{12}$

3. $\frac{8}{9} - \frac{3}{9}$

4. $\frac{15}{26} - \frac{7}{26}$

2·3 ADDITION AND SUBTRACTION

Adding and Subtracting Fractions with Unlike Denominators

You can use models to add fractions with unlike denominators.

$\frac{1}{2}$	$\frac{1}{2}$

$\frac{1}{3}$	$\frac{1}{3}$	$\frac{1}{3}$

$\frac{1}{6}$	$\frac{1}{6}$	$\frac{1}{6}$	$\frac{1}{6}$	$\frac{1}{6}$	$\frac{1}{6}$

$$\frac{1}{2} + \frac{1}{3} = \frac{5}{6}$$

To add or subtract fractions with unlike denominators, you need to change the fractions to equivalent fractions with common, or like, denominators before you find the sum or difference.

> ### ADDING AND SUBTRACTING FRACTIONS WITH UNLIKE DENOMINATORS
>
> Add $\frac{2}{3} + \frac{3}{4}$.
>
> - Find the *least common denominator* (p. 104) of the fractions.
>
>   12 is the LCD of 3 and 4.
>
> - Write equivalent fractions with the LCD.
>
> $\frac{2}{3} = \frac{2}{3} \times \frac{4}{4} = \frac{8}{12}$ and $\frac{3}{4} = \frac{3}{4} \times \frac{3}{3} = \frac{9}{12}$
>
> - Add or subtract the numerators. Put the result over the common denominator.
>
> $\frac{8}{12} + \frac{9}{12} = \frac{17}{12}$
>
> - Simplify, if possible.
>
> $\frac{17}{12} = 1\frac{5}{12}$

Check It Out

Add or subtract. Simplify, if possible.

5. $\frac{3}{8} + \frac{3}{4}$ 　　　　　6. $\frac{5}{6} - \frac{3}{5}$

Adding and Subtracting Mixed Numbers

Adding and subtracting mixed numbers is similar to adding and subtracting fractions. To subtract, sometimes you have to rename your number. Sometimes you will have to simplify an improper fraction in the answer.

Adding Mixed Numbers with Common Denominators

To add mixed numbers with common denominators, you need to write the sum of the numerators over the common denominator. Then add the whole numbers.

ADDING MIXED NUMBERS WITH COMMON DENOMINATORS

Add $7\frac{2}{5} + 6\frac{4}{5}$.

Add the whole numbers. $\left.\begin{array}{r} 7\frac{2}{5} \\ + 6\frac{4}{5} \end{array}\right\}$ Add the fractions.

$$\overline{13\frac{6}{5}}$$

Simplify if possible. $13\frac{6}{5} = 14\frac{1}{5}$

 Check It Out

Add or subtract. Simplify if possible.

7. $4\frac{2}{5} + 1\frac{3}{5}$

8. $12\frac{3}{8} - 2\frac{1}{8}$

9. $24\frac{5}{8} - 19\frac{3}{8}$

Adding Mixed Numbers with Unlike Denominators

You can use fraction drawings to model the addition of mixed numbers with unlike denominators.

$$1\frac{1}{2}$$
$$+\ 1\frac{1}{3}$$
$$2\frac{5}{6}$$

To add mixed numbers with unlike denominators you need to write equivalent fractions with a common denominator.

ADDING MIXED NUMBERS WITH UNLIKE DENOMINATORS

Add $2\frac{2}{3} + 4\frac{1}{2}$.

- Write the fractions with a common denominator.

  $\frac{2}{3} = \frac{4}{6}$ and $\frac{1}{2} = \frac{3}{6}$

- Then add and simplify.

Add the whole numbers. $\left.\begin{array}{r} 2\frac{4}{6} \\ +\ 4\frac{3}{6} \end{array}\right\}$ Add the fractions.

$$6\frac{7}{6}$$

Simplify, if possible. $6\frac{7}{6} = 7\frac{1}{6}$

$$2\frac{2}{3} + 4\frac{1}{2} = 7\frac{1}{6}$$

Check It Out

Add. Simplify, if possible.

10. $3\frac{5}{9} + 4\frac{1}{6}$

11. $4\frac{3}{5} + 5\frac{11}{15}$

Subtracting Mixed Numbers

You can model the subtraction of mixed numbers with unlike denominators.

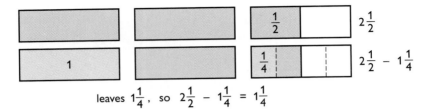

leaves $1\frac{1}{4}$, so $2\frac{1}{2} - 1\frac{1}{4} = 1\frac{1}{4}$

To subtract mixed numbers, you need to have or make common denominators.

SUBTRACTING MIXED NUMBERS

Subtract $9\frac{1}{5} - 4\frac{2}{3}$.

- If you have unlike denominators, write equivalent fractions with a common denominator.

  $9\frac{1}{5} = 9\frac{3}{15}$ and $4\frac{2}{3} = 4\frac{10}{15}$

- Subtract and simplify.

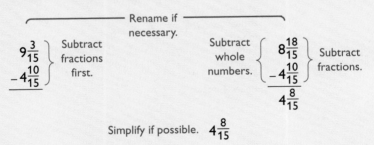

$9\frac{1}{5} - 4\frac{2}{3} = 4\frac{8}{15}$

Check It Out

12. $8\frac{1}{2} - 4\frac{2}{3}$ 13. $7\frac{1}{10} - 4\frac{3}{5}$

2·3 EXERCISES

Add or subtract.

1. $\frac{5}{7} + \frac{6}{11}$

2. $\frac{1}{4} + \frac{1}{3}$

3. $\frac{2}{5} - \frac{1}{10}$

4. $\frac{7}{8} - \frac{3}{4}$

5. $\frac{7}{12} - \frac{3}{10}$

6. $\frac{3}{10} + \frac{4}{5}$

7. $5\frac{2}{3} + 2\frac{1}{2}$

8. $4\frac{1}{2} - 3\frac{3}{4}$

9. $1\frac{2}{3} + 1\frac{1}{4}$

10. $7\frac{3}{5} - 3\frac{2}{3}$

11. $2\frac{3}{4} + 6\frac{5}{16}$

12. $9\frac{4}{7} - 5\frac{1}{14}$

13. $8\frac{3}{8} + 6\frac{3}{4}$

14. $10\frac{1}{10} - 3\frac{3}{20}$

15. $9\frac{1}{2} - 4\frac{7}{8}$

16. Desrie is planning to make a two-piece costume. One piece requires $1\frac{5}{8}$ yd of material and the other requires $1\frac{3}{4}$ yd. She has $4\frac{1}{2}$ yd of material. Does she have enough to make the costume?

17. Atiba was $53\frac{7}{8}$ in. tall on his birthday last year. On his birthday this year he was $56\frac{1}{4}$ in. tall. How much did he grow during the year?

18. Hadas's punch bowl holds 8 qt. Can she serve cranberry punch made with $6\frac{2}{3}$ qt cranberry juice and $2\frac{1}{4}$ qt apple juice?

19. Nia jogged $4\frac{1}{10}$ mi on Sunday, $2\frac{2}{5}$ mi on Tuesday, and $3\frac{1}{2}$ mi on Thursday. How many miles does she have to jog on Saturday to reach her weekly goal of $12\frac{1}{2}$ mi?

2·4 Multiplication and Division of Fractions

Multiplying Fractions

You know that 3×2 means "3 groups of 2." Multiplying fractions involves the same concept: $3 \times \frac{1}{4}$ means "3 groups of $\frac{1}{4}$." You will find it helpful to think of *times* as meaning *of*.

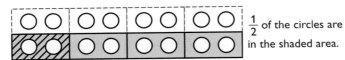

1 group of $\frac{1}{4}$ →

3 groups of $\frac{1}{4}$ or $\frac{3}{4}$

The same is true when you are multiplying a fraction by a fraction. For example, $\frac{1}{4} \times \frac{1}{2}$ means you would actually be finding $\frac{1}{4}$ of $\frac{1}{2}$.

$\frac{1}{2}$ of the circles are in the shaded area.

$\frac{1}{4}$ of $\frac{1}{2}$ of the circles are $\frac{1}{8}$ of *all* the circles.   So $\frac{1}{4} \times \frac{1}{2} = \frac{1}{8}$.

When you are not using models to multiply fractions, you multiply the numerators and then the denominators. There's no need to find a common denominator.

MULTIPLYING FRACTIONS

Multiply $\frac{2}{3} \times 2\frac{1}{5}$.

- Convert mixed numbers, if any, to *improper fractions* (p. 107).

$$\frac{2}{3} \times 2\frac{1}{5} = \frac{2}{3} \times \frac{11}{5}$$

- Multiply the numerators and the denominators.

$$\frac{2}{3} \times \frac{11}{5} = \frac{2 \times 11}{3 \times 5} = \frac{22}{15}$$

- Write the **products** in *lowest terms* (p. 106).

$$\frac{22}{15} = 1\frac{7}{15}$$

$$\frac{2}{3} \times 2\frac{1}{5} = 1\frac{7}{15}$$

Check It Out

1. $\frac{1}{2} \times \frac{3}{5}$ 2. $3\frac{1}{3} \times \frac{1}{2}$

Shortcut for Multiplying Fractions

You can use a shortcut when you multiply fractions. Instead of multiplying across and then writing the product in lowest terms, you can cancel **factors** first.

CANCELING FACTORS

Multiply $\frac{4}{5} \times \frac{15}{16}$.

- Write mixed numbers, if any, as improper fractions.

- Cancel factors, if you can.

- Multiply.

- Write the products in lowest terms, if necessary.

$$\frac{4}{5} \times \frac{15}{16}$$

$$= \frac{\overset{1}{\cancel{4}}}{\cancel{5}_{1}} \times \frac{\overset{1}{\cancel{5}} \times 3}{\cancel{4}_{1} \times 4}$$

$$= \frac{1}{1} \times \frac{3}{4}$$

$$= \frac{3}{4}$$

 Check It Out

3. $\frac{4}{7} \times \frac{21}{24}$ 4. $\frac{3}{5} \times \frac{20}{21}$ 5. $3\frac{1}{5} \times 1\frac{1}{4}$

Finding the Reciprocal of a Number

To find the **reciprocal** of a number, you switch the numerator and the denominator.

Number	$\frac{3}{5}$	$2 = \frac{2}{1}$	$3\frac{1}{2} = \frac{7}{2}$
Reciprocal	$\frac{5}{3}$	$\frac{1}{2}$	$\frac{2}{7}$

When you multiply a number by its reciprocal, the product is 1.

$$\frac{3}{8} \times \frac{8}{3} = \frac{24}{24} = 1$$

The number 0 does not have a reciprocal.

 Check It Out

Find the reciprocal of each number.

6. $\frac{2}{5}$ 7. 4   8. $2\frac{1}{3}$

2·4 MULTIPLICATION AND DIVISION

Dividing Fractions

When you divide a fraction by a fraction, such as $\frac{1}{2} \div \frac{1}{6}$, you are really finding out how many $\frac{1}{6}$'s are in $\frac{1}{2}$. That's why the answer is 3. To divide fractions, you replace the divisor with its reciprocal and then multiply to get your answer.

$$\frac{1}{2} \div \frac{1}{6} = \frac{1}{2} \times \frac{6}{1} = 3$$

DIVIDING FRACTIONS

Divide $\frac{4}{5} \div 2\frac{2}{3}$.

- Write any mixed number as improper fractions.

$$\frac{4}{5} \div \frac{8}{3}$$

- Replace the divisor with its reciprocal and cancel factors.

$$\frac{4}{5} \times \frac{3}{8} = \frac{\overset{1}{4}}{5} \times \frac{3}{\underset{2}{8}} = \frac{1}{5} \times \frac{3}{2}$$

- Multiply.

$$\frac{1}{5} \times \frac{3}{2} = \frac{3}{10}$$

Check It Out

9. $\frac{2}{3} \div \frac{1}{8}$ 10. $\frac{1}{4} \div \frac{1}{14}$ 11. $\frac{3}{10} \div \frac{2}{5}$

Oseola McCarty

Miss Oseola McCarty had to leave school after sixth grade. At first, she charged $1.50 to do a bundle of laundry, later $10.00. But she always managed to save. By age 86, she had accumulated $250,000. In 1995, she decided to donate $150,000 to endow a scholarship. Miss McCarty said, "The secret to building a fortune is compounding interest. You've got to leave your investment alone long enough for it to increase."

2·4 EXERCISES

Multiply.

1. $\frac{2}{3} \times \frac{7}{8}$

2. $\frac{1}{8} \times 2\frac{1}{2}$

3. $3\frac{1}{3} \times \frac{3}{5}$

4. $\frac{4}{9} \times 1\frac{1}{2}$

5. $\frac{6}{8} \times 3\frac{1}{4}$

6. $2\frac{1}{2} \times 1\frac{1}{3}$

7. $\frac{7}{8} \times 1\frac{2}{7}$

8. $1\frac{1}{3} \times 1\frac{1}{2}$

9. $3\frac{7}{8} \times \frac{9}{12}$

10. $\frac{3}{4} \times \frac{6}{11}$

11. $6\frac{3}{9} \times 2\frac{8}{14}$

12. $\frac{4}{5} \times \frac{6}{7}$

Find the reciprocal of each number.

13. $\frac{4}{5}$

14. 3

15. $4\frac{1}{2}$

16. $4\frac{2}{3}$

17. $\frac{6}{7}$

Divide.

18. $\frac{2}{3} \div \frac{1}{2}$

19. $\frac{5}{8} \div \frac{3}{4}$

20. $1\frac{3}{4} \div \frac{5}{6}$

21. $2\frac{1}{3} \div \frac{2}{3}$

22. $\frac{7}{15} \div \frac{1}{6}$

23. $1\frac{1}{7} \div \frac{2}{5}$

24. $\frac{3}{16} \div \frac{1}{4}$

25. $2\frac{1}{3} \div 9$

26. $5\frac{1}{5} \div \frac{1}{2}$

27. $3\frac{1}{2} \div 2\frac{1}{3}$

28. $2\frac{9}{24} \div 3\frac{8}{12}$

29. Naoko's spaghetti sauce recipe calls for $1\frac{1}{3}$ tsp of salt for every quart of sauce. How many teaspoons of salt will she need to make $5\frac{1}{2}$ qt of sauce?

30. Aisha plans to buy corduroy material at $4 per yd. If she needs to buy $8\frac{1}{2}$ yd of material, how much will it cost?

2·5 Naming and Ordering Decimals

Decimal Place Value: Tenths and Hundredths

You can use what you know about **place value** for **whole numbers** to help you read and write decimals.

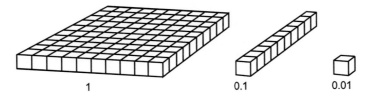

The base-ten blocks show that:
One whole (1) is 10 times greater than 1 tenth (0.1).
One tenth (0.1) is ten times greater than 1 hundredth (0.01).

You can use a place-value chart to help you read and write decimal numbers.

ten thousands	thousands	hundreds	tens	ones	tenths	hundredths	
			3	5	2	6	thirty-five and twenty-six hundredths
	1	0	2	0	0	2	one thousand twenty and two hundredths
			7	0	7		seventy and seven tenths
			7	0	7	0	seventy and seventy hundredths
				0	3		three tenths

To read the decimal, read the whole number to the left of the decimal point as usual. You say "and" for the decimal point. Then find the place of the last decimal digit and use it to name the decimal.

To write a decimal, you can write the whole number, put a decimal point, then place the last digit of the decimal number in the place that names it.

1,000 + 20 + .02 is 1,020.02 written in expanded notation. The place-value chart can help you write expanded notation. You write each nonzero place as a number and add them together.

Check It Out
Write the decimal.
1. sixth tenths
2. forty-four hundredths
3. eight and seventeen hundredths
4. three and four tenths

Decimal Place Value: Thousandths

Thousandths is used as an accurate measurement in sports statistics and scientific studies. The number 1 is 1,000 times one thousandth. 1 hundredth is equal to 10 thousandths. $(\frac{1}{100} = \frac{10}{1,000})$.

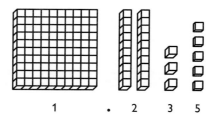

1	. 2	3	5

The base ten blocks model the decimal number 1.235. You read 1.235 "one *and* two hundred thirty-five thousandths." 1.235 in expanded form is 1 + 0.2 + 0.03 + 0.005

Check It Out
Write the decimal in words.
5. 0.382
6. 2.154
7. 0.066

Naming Decimals Greater Than and Less Than One

Decimal numbers are based on units of ten.

The chart shows the place value for some of the digits of a decimal. You can use a place-value chart to help you name decimals greater than and less than one.

NAMING DECIMAL NUMBERS

Read 27.6314.

- Values to the left of the decimal point are greater than one.

  27 means 2 tens and 7 ones.

- Read the decimal. The word name of the decimal is determined by the place value of the digit in the last place.

  The last digit (4) is in the ten-thousandth place.

27.6314 is read as twenty-seven *and* six thousand three hundred fourteen ten thousandths.

Check It Out

Use the place-value chart to tell what each boldfaced digit means. Then write the numbers in words.

8. 4.4**1**1
9. 0.0**3**2
10. 5.00**4**6
11. 0.00**3**41

Comparing Decimals

When zeros are added to the right of the decimal in the following manner, they do not change the value of the number.

1.039 = 1.0390 = 1.03900 = 1.039000...

To compare decimals, you compare their place values.

COMPARING DECIMALS

Compare 19.5032 and 19.5047.

- Start at the left. Find the first place where the numbers are different.

$$1\,9\,.\,5\,0\,\boxed{3}\,2$$
$$1\,9\,.\,5\,0\,\boxed{4}\,7$$
↑
The thousandths place is different.

- Compare the digits that are different.

3 < 4

- The numbers compare the same way the digits compare.
19.5032 < 19.5047

Check It Out

Compare. Write >, <, or =.

12. 26.3 ☐ 26.4   13. 0.0176 ☐ 0.0071

Ordering Decimals

To write decimals from least to greatest and vice versa, you need to first compare the numbers two at a time.

Order the decimals 1.143, 0.143, and 1.14.
- Compare the numbers two at a time.
  1.143 > 1.140   1.140 > 0.143
- List the decimals from least to greatest.
  0.143; 1.14; 1.143

Check It Out
Write in order from least to greatest.
14. 3.0186; 30.618; 3.1608
15. 9.083; 9.381; 93.8; 9.084; 9
16. 0.622; 0.662; 0.6212; 0.6612

Rounding Decimals

Rounding decimals is similar to rounding whole numbers.

Round 14.046 to the nearest hundredth.
- Find the rounding place.
  14.046
  ↑
  hundredths
- Look at the digit to the right of the rounding place.
  14.0**4**6
  If it is less than 5, leave the digit in the rounding place unchanged. If it is more than or equal to 5, increase the digit in the rounding place by 1.
  6 > 5
- Write the rounded number.
  14.046 rounded to the nearest hundredth is 14.05.

Check It Out
Round each decimal to the nearest hundredth.
17. 1.544   18. 36.389
19. 8.299   20. 8.681

2·5 EXERCISES

Write the decimal.
1. two and forty-six hundredths
2. eight tenths
3. nine hundred thirty-six ten thousandths
4. twenty-five and two hundred ten hundred thousandths

Write the decimal in words.
5. 0.98
6. 78.122
7. 0.4444

Give the value of each boldfaced digit.
8. 34.24**1**
9. **4**.3461
10. 0.129**6**
11. 24.**1**4

Compare. Use <, >, or =.
12. 15.099 ☐ 15.11
13. 12.5640 ☐ 12.56
14. 7.1498 ☐ 7.2
15. 0.684 ☐ 0.694

List in order from least to greatest.
16. 0.909; 0.090; 0.90; 0.999
17. 8.822; 8.288; 8.282; 8.812
18. 6.85; 0.68; 0.685; 68.5

Round each decimal to the indicated place value.
19. 1.6432, thousandths
20. 48.098, hundredths
21. 3.86739, ten thousandths

22. Four ice skaters are in a competition in which the highest possible score is 10.0. Three of the skaters have completed their performances and their scores are 9.61, 9.65, and 9.60. What score must the last skater get in order to win the competition?

23. The Morales family traveled 45.66 mi on Saturday, 45.06 mi on Sunday, and 45.65 mi on Monday. On which day did they travel the farthest?

2·6 Decimal Operations

Adding and Subtracting Decimals

Adding and subtracting decimals is similar to adding whole numbers. You need to be careful to line up the appropriate digits.

ADDING AND SUBTRACTING DECIMALS

Add 4.75 + 0.6 + 32.46.

- Line up the decimal points.

$$\begin{array}{r} 4.75 \\ .6 \\ + 32.46 \\ \hline \end{array}$$

- Add (or subtract) the place farthest right. Carry or borrow as necessary.

$$\begin{array}{r} {}^{1} \\ 4.75 \\ .6 \\ + 32.46 \\ \hline 1 \end{array}$$

- Add (or subtract) the next place left. Carry or borrow as necessary.

$$\begin{array}{r} {}^{1\,1} \\ 4.75 \\ 0.6 \\ + 32.46 \\ \hline 81 \end{array}$$

- Continue through the whole numbers. Place the decimal point in the result.

$$\begin{array}{r} 4.75 \\ 0.6 \\ + 32.46 \\ \hline 37.81 \end{array}$$

 Check It Out

Solve.
1. 8.1 + 31.75
2. 19.58 + 37.42 + 25.75
3. 17.8 − 4.69
4. 52.7 − 0.07219

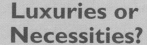

Luxuries or Necessities?

Recent economic reforms have made China one of the fastest growing economies in the world. After years of hardship, the country hasn't yet caught up in providing luxuries for its vast population of approximately 1,200,000,000.

Standard-of-living refers to the level of goods, services, and luxuries available to an individual or a population. Here are two examples of how China's standard-of-living compares with ours.

	China	United States
Number of people per telephone	36.4	1.3
Number of people per TV	6.7	1.2

When China has the same number of telephones per person as the United States has, how many will it have? See Hot Solutions for answer.

2·6 DECIMAL OPERATIONS

Estimating Decimal Sums and Differences

One way that you can **estimate** decimal sums and differences is to use compatible numbers. Compatible numbers are close to the numbers in the problem and are easy to work with mentally.

Estimate the sum of 2.244 + 6.711.
• Replace the numbers with compatible numbers.

$$2.244 \rightarrow 2$$
$$6.711 \rightarrow 7$$

• Add the numbers.

$$2 + 7 = 9$$

Estimate the difference of 12.6 − 8.4.

$$12.6 \rightarrow 13$$
$$8.4 \rightarrow 8$$
$$13 - 8 = 5$$

Check It Out
Estimate each sum or difference.
5. 8.64 + 5.33
6. 11.3 − 9.4
7. 18.145 − 3.66
8. 3.48 + 5.14 + 8.53

Multiplying Decimals

Multiplying decimals is much the same as multiplying whole numbers. You can model the multiplication of decimals with a 10 × 10 grid. Each tiny square is equal to one hundredth.

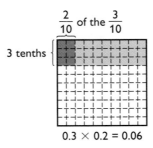

$\frac{2}{10}$ of the $\frac{3}{10}$

3 tenths

0.3 × 0.2 = 0.06

MULTIPLYING DECIMALS

Multiply 32.6 × 0.08.

• Multiply as with whole numbers.

32.6	326
× 0.08	× 8
	2608

• Add the number of decimal places for the factors.

32.6 → 1 decimal place

0.08 → 2 decimal places

1 + 2 = 3 total decimal places

• Place the decimal point in the product.

32.6 → 1 decimal place

× 0.08 → 2 decimal places

2.608 → 3 decimal places

32.6 × 0.08 = 2.608

Check It Out

Multiply.

9. 10.9 × 0.7

10. 6.2 × 0.087

11. 0.61 × 3.2

12. 9.81 × 6.5

2·6 DECIMAL OPERATIONS

Multiplying Decimals with Zeros in the Product

Sometimes when you are multiplying decimals, you need to add zeros in the product.

Multiply 0.7 × 0.0345.

- Multiply as with whole numbers.

$$
\begin{array}{r} 0.0345 \\ \times\ \ 0.7 \end{array}
\qquad
\begin{array}{r} 00345 \\ \times\ \ 07 \\ \hline 2415 \end{array}
$$

- Count the number of decimal places in the factors.

$$
\begin{array}{r} 00345 \\ \times\ \ 07 \\ \hline 2415 \end{array}
\quad
\begin{array}{l} \longrightarrow \ \ 4\ \text{decimal places} \\ \longrightarrow \ \ 1\ \text{decimal place} \\ \longrightarrow \ \ \text{needs 5 decimal places} \end{array}
$$

- Add zeros in the product as necessary.

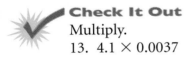

.02415

   Since 5 places are needed in the product, write a zero to the left of the 2.

0.7 × 0.0345 = 0.02415

Check It Out
Multiply.

13. 4.1 × 0.0037   14. 0.961 × 0.05

Estimating Decimal Products

To estimate decimal products, you replace given numbers with compatible numbers. Compatible numbers are estimates you choose because they are easier to work with mentally.

To estimate the product of 27.3 × 44.2, you start by replacing the factors with compatible numbers. 27.3 becomes 30. 44.2 becomes 40. Then multiply mentally. 30 × 40 = 1,200. So 27.3 × 44.2 is about 1,200.

Check It Out

Estimate each product using compatible numbers.
15. 25.71×9.4 16. 9.48×10.73

Dividing Decimals

Dividing decimals is similar to dividing whole numbers. You can use a model to help you understand dividing decimals. For example, $0.9 \div 0.3$ means how many groups of 0.3 are in 0.9?

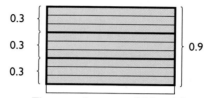

There are 3 groups of 0.3 in 0.9.

There are 3 groups of 0.3 in 0.9, so $0.9 \div 0.3 = 3$.

DIVIDING DECIMALS

Divide $31.79 \div 1.1$.

- Multiply the divisor by a power of ten to make it a whole number.

$$1.1\overline{)31.79} \qquad 1.1 \times 10 = 11$$

- Multiply the dividend by the same power of ten.

$$31.79 \times 10 = 317.9 \qquad 11.\overline{)317.9}$$

- Place the decimal point in the quotient.

$$11.\overline{)317.9}^{\cdot}$$

- Divide.

$$\overset{28.9}{11.\overline{)317.9}}$$

Check It Out
Divide.
17. $0.231 \div 0.07$
18. $0.312 \div 0.06$
19. $1.22 \div 0.4$
20. $0.6497 \div 8.9$

Rounding Decimal Quotients

You can use a calculator to divide decimals and round quotients.

Divide $6.3 \div 2.6$. Round the quotient to the nearest hundredth.

- Use your calculator to divide.

  6.3 ÷ 2.6 = $\boxed{2.4230769}$

- To round the quotient, look at one place to the right of the rounding place. If the digit to the right of the rounding place is 5 or above, round up. If the digit to the right of the rounding place is less than 5, the digit to be rounded remains the same.

  2.4230769 $3 < 5$, so 2.4230769 rounded to the nearest hundredth is 2.42.

$6.3 \div 2.6 = 2.42$

Check It Out
Use a calculator to find each quotient. Round to the nearest hundredth.
21. $2.2 \div 0.3$
22. $32.5 \div 0.32$
23. $0.671 \div 2.33$

2·6 EXERCISES

Estimate each sum or difference.
1. 3.24 + 1.06
2. 6.09 − 3.7
3. 8.445 + 0.92
4. 3.972 + 4.124
5. 11.92 − 8.0014

Add.
6. 234.1 + 4.92
7. 65.11 + 22.64
8. $11.19 + $228.16
9. 7.0325 + 0.81
10. 1.8 + 4 + 2.6473

Subtract.
11. 22 − 11.788
12. 42.108 − 0.843
13. 386.1 − 2.94
14. 52.12 − 18.666
15. 12.65 − 3.0045

Multiply.
16. 0.7 × 6.633
17. 12.6 × 33.44
18. 0.14 × 0.02
19. 49.32 × 0.6484
20. 0.57 × 0.91

Divide.
21. 43.68 ÷ 5.2
22. 6.552 ÷ 9.1
23. 65.026 ÷ 0.82
24. 2.175 ÷ 2.9

Divide. Round the quotient to the nearest hundredth.
25. 18.47 ÷ 5.96
26. 18.6 ÷ 2.8
27. 82.3 ÷ 8.76
28. 63.7 ÷ 7.6

29. Bananas at the Quick Stop Market cost $0.49 per lb. Find the cost of 4.6 lb of bananas. Round your answer to the nearest cent.

30. How much change would you get from a $50 bill if you bought two $14.95 shirts?

2·6 EXERCISES

2·7 Meaning of Percent

Naming Percents

A **ratio** of a number to 100 is called a **percent.** Percent means *per hundred* and is represented by the symbol %.

You can use graph paper to model percents. There are 100 squares on a 10 by 10 sheet of graph paper. So a 10 by 10 sheet can be used to represent 100%. Because percent means how many out of 100, it is easy to tell what percent of the 100-square graph paper is shaded.

25 of 100 are blue
(25% blue).

50 of 100 are white
(50% white).

10 of 100 are red
(10% red).

15 of 100 are yellow
(15% yellow).

 Check It Out

What percent of each square is shaded?

1.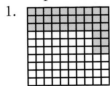

2.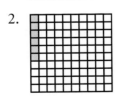

3.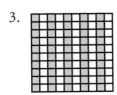

Understanding the Meaning of Percent

One way to think about percents is to become very comfortable with a few **benchmarks.** You build what you know about percents based on these different benchmarks. You can use these benchmarks to help you estimate percents of other things.

none			half		all
0 $\frac{1}{100}$ $\frac{1}{10}$	$\frac{1}{4}$		$\frac{1}{2}$	$\frac{3}{4}$	100%
0.01 0.10	0.25		0.50	0.75	
1% 10%	25%		50%	75%	

ESTIMATING PERCENTS

Estimate 48% of 50.

- Choose a benchmark, or combination of benchmarks, close to the target percent.

  48% is close to 50%.

- Find the fraction or decimal equivalent to the benchmark percent.

  50% $= \frac{1}{2}$

- Use the benchmark equivalent to estimate the percent.

  $\frac{1}{2}$ of 50 is 25.

So 48% of 50 is about 25.

Check It Out

Use fractional benchmarks to estimate the percents.

4. 49% of 100   5. 24% of 100
6. 76% of 100   7. 11% of 100

2·7 MEANING OF PERCENT

Using Mental Math to Estimate Percent

To help you estimate the percent of something, such as a tip in a restaurant, you can use fraction or decimal benchmarks in real-life situations.

Estimate a 10% tip for a bill of $4.48.
- Round to a convenient number.
    $4.48 rounds to $4.50.
- Think of the percent as a benchmark or combination of benchmarks.
    $10\% = 0.10 = \frac{1}{10}$
- Multiply mentally. Combine, if necessary.
    $0.10 \times \$4.50 = \$0.45$ The tip is about $0.45.

 **Check It Out**

Estimate the amount of each tip.
8. 10% of $2.50   9. 20% of $85

Honesty Pays

David Hacker, a cabdriver, found a wallet in the back seat of his cab that contained $25,000—about a year's salary for him.

The owner's name was in the wallet and Hacker remembered where he had dropped him off. He went straight to the hotel and found the man. The owner, a businessman, had already realized he'd lost his wallet and figured he'd never see it again. He didn't believe anyone would be that honest! On the spot, he handed the cabdriver fifty $100 bills.

What percent of the money did Hacker receive as a reward? See Hot Solutions for answer.

2·7 MEANING OF PERCENT

2·7 EXERCISES

Write the percent that is shaded.

1.

2.

3.

4.

5.

6.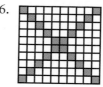

Use fractional benchmarks to estimate the percent of each number.

7. 26% of 80

8. 45% of 200

9. 12% of 38

10. 81% of 120

11. Estimate a 15% tip for a $60 bill.

12. Estimate a 20% tip for a $39 bill.

13. Estimate a 10% tip for a $84 bill.

14. Estimate a 25% tip for a $75 bill.

15. Estimate a 6% tip for a $55 bill.

2·8 Using and Finding Percents

Finding a Percent of a Number

To find the percent of a number, you can use decimals or fractions. You first change the percent to a decimal or a fraction. Sometimes it is easier to change to a decimal representation and other times to a fractional one.

To find 25% of 60, you can use either the fraction method or the decimal method.

FINDING THE PERCENT OF A NUMBER: TWO METHODS

Find 25% of 60.

DECIMAL METHOD

- Change percent to decimal.

$$25\% = 0.25$$

- Multiply.

$$0.25 \times 60 = 15$$

So 25% of 60 = 15.

PERCENT METHOD

- Change percent to fraction.

$$25\% = \frac{25}{100}$$

- Reduce fraction to lowest terms.

$$\frac{25}{100} = \frac{1}{4}$$

- Multiply.

$$\frac{1}{4} \times \frac{60}{1} = \frac{60}{4} = 15$$

So 25% of 60 = 15.

Check It Out

Find the percent of each number.

1. 65% of 80
2. 85% of 500
3. 30% of 90
4. 75% of 420

Finding the Percent of a Number: Proportion Method

You can use proportions to help you find the percent of a number.

FINDING THE PERCENT OF A NUMBER: PROPORTION METHOD

Gwen works in a music store. She receives a commission of 10% on her sales. Last month she sold $850 worth of CDs. What was her commission?

- Use a proportion to find the percent of a number.

P = Part (of the base or total) $\qquad \frac{P}{R} = \frac{B}{100}$

B = Base (total) $\quad R$ = Rate (percentage)

- Ientify the given items before trying to find the unknown.

P is unknown, call x. $\quad R$ is 10%. $\quad B$ is $850.

- Set up the proportion.

$\frac{P}{R} = \frac{B}{100}$ $\qquad\qquad \frac{x}{10} = \frac{850}{100}$

- Cross multiply.

$100x = 8,500$

- Divide both sides of the equation by the coefficient of x.

$\frac{8,500}{100} = \frac{100x}{100}$ $\qquad\qquad 85 = x$

Gwen received a commission of $85.

Check It Out

Use a proportion to find the percent of each number.

5. 76% of 39
6. 14% of 85
7. 66% of 122
8. 55% of 300

2·8 USING AND FINDING PERCENTS

Finding Percent and Base

Setting up and solving a proportion can help you find what percent a number is of a second number. Use the ratio $\frac{P}{B} = \frac{R}{100}$ where $P =$ Part (of base), $B =$ Base (total), and $R =$ Rate (percentage).

FINDING THE PERCENT

What percent of 40 is 15?
- Set up a proportion, using this form.

$$\frac{\text{Part}}{\text{Base}} = \frac{\text{Percent}}{100}$$
$$\frac{15}{40} = \frac{n}{100}$$

(The number after the word *of* is the base.)
- Find the cross products of the proportion.

$$100 \times 15 = 40 \times n$$
- Find the products.

$$1,500 = 40n$$
- Divide both sides of the equation by the *coefficient* of n.

$$\frac{1,500}{40} = \frac{40n}{40} \qquad \frac{1,500}{40} = n \qquad 37.5\% = n$$

15 is 37.5% of 40.

Check It Out

Solve.

9. What percent of 240 is 80?
10. What percent of 64 is 288?
11. What percent of 2 is 8?
12. What percent of 55 is 33?

FINDING THE BASE

8 is 32% of what number?

- Set up a percent proportion using this form.

$$\frac{Part}{Base} = \frac{Percent}{100}$$

$$\frac{8}{n} = \frac{32}{100}$$

  (The phrase *what number* after the word *of* is the base.)

- Find the cross products of the proportion.

$$8 \times 100 = 32 \times n$$

- Find the products.

$$800 = 32n$$

- Divide both sides of the equation by the coefficient of *n*.

$$\frac{800}{32} = \frac{32n}{32}$$

$$n = 25$$

8 is 32% of 25.

 Check It Out

13. 52 is 50% of what number?
14. 15 is 75% of what number?
15. 40 is 160% of what number?
16. 84 is 7% of what number?

Percent of Increase or Decrease

Sometimes it is helpful to keep a record of your monthly expenses. Keeping a record allows you to see the actual *percent of increase* or *decrease* in your expenses. You can make a chart to record your expenses.

2·8 USING AND FINDING PERCENTS

Expenses	November	December	Amount of Increase or Decrease	% of Increase or Decrease
Clothing	$125	$ 76	$49	39%
Entertainment	$ 44	$ 85		
Food	$210	$199	$11	5%
Miscellaneous	$110	$ 95		
Rent	$450	$465	$15	3%
Travel	$ 83	$ 65		
Total	$1,022	$985	$37	4%

You can use a calculator to find the percent of increase or decrease.

FINDING THE PERCENT OF INCREASE

During November $44 was spent on entertainment. In December the amount spent on entertainment was $85.

- Use a calculator to key in the following:

$$\underset{\text{amount}}{\text{new}} \;\boxed{-}\; \underset{\text{amount}}{\text{original}} \;\boxed{=}\; \underset{\text{of increase}}{\text{amount}}$$

$$85\;\boxed{-}\;44\;\boxed{=}\;\boxed{\qquad 41.\quad}$$

- Leave the amount of increase on display.

$$\boxed{\qquad 41 \qquad}$$

- Use your calculator to divide the amount of increase by the original amount.

$$\underset{\text{increase}}{\text{amount of}} \;\boxed{\div}\; \underset{\text{amount}}{\text{original}} \;\boxed{=}\; \underset{\text{increase}}{\text{percent of}}$$

$$41\;\boxed{\div}\;44\;\boxed{=}\;\boxed{\qquad .9318182\quad}$$

Round to nearest hundredth and convert to percent.

$$0.9318182 = 0.93 = 93\%$$

The percent of increase from $44 to $85 is 93%.

FINDING THE PERCENT OF DECREASE

During November, $83 was spent on travel. In December, $65 was spent on travel. You can use a calculator to find the percent of decrease.

- Use a calculator to key in the following:

| original amount | $\boxed{-}$ | new amount | $\boxed{=}$ | amount of decrease |

83 $\boxed{-}$ 65 $\boxed{=}$ $\boxed{\qquad 18.}$

- Leave the amount of decrease on display.

$\boxed{\qquad 18.}$

- Use your calculator to divide the amount of decrease by the original amount.

| amount of decrease | $\boxed{\div}$ | original amount | $\boxed{=}$ | percent of decrease |

$\boxed{\qquad 18.}$ $\boxed{\div}$ 83 $\boxed{=}$ $\boxed{0.2168674}$

Round and convert from decimal to percent.

0.216874 = 0.22 = 22%

The percent of decrease from $83 to $65 is 22%.

Check It Out

Use a calculator to find the percent of increase or decrease.

17. 16 to 38   18. 43 to 5
19. 99 to 3   20. 6 to 10

Discounts and Sale Prices

A **discount** is the amount that an item is reduced from the regular price. The sale price is the regular price minus the discount. Discount stores have regular prices below the suggested retail price. You can use percents to find discounts and resulting sale prices.

2·8 USING AND FINDING PERCENTS

This CD player has a regular price of $109.99. It is on sale for 25% off the regular price. How much money will you save by buying the item on sale?

25% OFF

CD Player

You can use a calculator to help you find the discount and resulting sale price of the item.

FINDING DISCOUNTS AND SALE PRICES

The regular price of an item is $109.99. It is marked 25% off. Find the discount and the sale price.

- Use a calculator to multiply the regular price times the discount price.

    regular price $\times$ discount percent $=$ discount

    109.99 $\times$ 25 $\%$ $=$ ⌐ 27.4975 ⌐

- If necessary, round the discount to the nearest hundredth.

    27.4975 = 27.50

    The discount is $27.50.

- Use a calculator to subtract the discount from the regular price. This will give you the sale price.

    regular price $-$ discount $=$ sale price

    109.99 $-$ 27.50 $=$ ⌐ 82.49 ⌐

The sale price is $82.49.

Check It Out

Use a calculator to find the discount and sale price.
21. Regular price: $90, Discount percent: 40%.
22. Regular price: $120, Discount percent: 15%.

Estimating a Percent of a Number

To estimate a percent of a number, you can use what you know about *compatible numbers* and simple fractions.

Percent	1%	5%	10%	20%	25%	33⅓%	50%	66⅔%	75%	100%
Fraction	$\frac{1}{100}$	$\frac{1}{20}$	$\frac{1}{10}$	$\frac{1}{5}$	$\frac{1}{4}$	$\frac{1}{3}$	$\frac{1}{2}$	$\frac{2}{3}$	$\frac{3}{4}$	1

The table can help you estimate the percent of a number.

ESTIMATING A PERCENT OF A NUMBER

Estimate 12% of 32.
- Find the percent that is closest to the percent you want to find.

  12% is about 10%.
- Find the fractional equivalent for the percent.

  10% is equivalent to $\frac{1}{10}$.
- Find a compatible number for the number you want to find the percent of.

  32 is about 30.
- Use the fraction to find the percent.

  $\frac{1}{10}$ of 30 is 3.

12% of 32 is about 3.

Check It Out

23. 19% of 112   24. 65% of 298

2·8 USING AND FINDING PERCENTS

Finding Simple Interest

When you have a savings account, the bank pays you for the use of your money. When you take out a loan from a bank, you pay the bank for the use of their money. In both situations, the money paid is called the *interest*. The amount of money you borrow or save is called the *principal*. To find out how much interest you will pay or earn, you can use the formula $I = P \times R \times T$. The chart below can help you to understand the formula.

P	Principal — the amount of money you borrow or save
R	Interest Rate — a percent of the principal you pay or earn
T	Time — the length of time you borrow or save
I	Total Interest — interest you pay or earn for the entire time
A	Amount — total amount (principal plus interest) you pay or earn

FINDING SIMPLE INTEREST

Use a calculator to find the total amount you will pay if you borrow $5,000 at 7% simple interest for 3 years.

- Multiply the principal (P) × the interest rate (R) × the time (T) to find the interest (I) you will pay.

  $P \times R \times T = I$

  5000 $\boxed{\times}$ 7 $\boxed{\%}$ $\boxed{\times}$ 3 $\boxed{=}$ $\boxed{\quad 1050.}$

  $1,050 is the interest.

- To find the total amount you will pay back, add the principal and the interest.

  $P + I = A$

  5000 $\boxed{+}$ 1050 $\boxed{=}$ $\boxed{\quad 6050.}$

$6,050 is the total amount of money to be paid back.

 Check It Out

Find the interest (I) and the total amount (A).

25. $P = \$750, R = 13\%, T = 2$ years
26. $P = \$3,600, R = 14\%, T = 9$ months

2·8 EXERCISES

Find the percent of each number.
1. 3% of 45
2. 44% of 125
3. 95% of 64
4. 2% of 15.4

Solve.
5. What percent of 40 is 29?
6. 15 is what percent of 60?
7. 4 is what percent of 18?
8. What percent of 5 is 3?
9. 64 is what percent of 120?

Solve.
10. 40% of what number is 30?
11. 25% of what number is 11?
12. 96% of what number is 24?
13. 67% of what number is 26.8?
14. 62% of what number is 15.5 ?

Find the percent of increase or decrease to the nearest percent.
15. 8 to 10
16. 45 to 18
17. 12 to 4
18. 15 to 20

Find the discount and sale price.
19. Regular price: $19.95, discount percent: 40%
20. Regular price: $65.99, discount percent: 15%
21. Regular price: $285.75, discount percent: 22%
22. Regular price: $385, discount percent: 40%

Find the interest and total amount. Use a calculator.
23. $P = \$8,500$, $R = 6.5\%$ per year, $T = 1$ year
24. $P = \$1,200$, $R = 7\%$ per year, $T = 2$ years
25. $P = \$2,400$, $R = 11\%$ per year, $T = 6$ months

Estimate the percent of each number.
26. 15% of 65
27. 27% of 74
28. 76% of 124

29. Lyudmila bought a CD player for 25% off the regular price of $179.99. How much did she save? How much did she pay?
30. A VCR is on sale for 15% off the regular price of $289.89. What is the discount? What is the sale price?

2·9 Fraction, Decimal, and Percent Relationships

Percents and Fractions

Percents and fractions both describe a ratio out of 100. The chart will help you to understand the relationship between percents and fractions.

Percent	Fraction
50 out of 100 = 50%	$\frac{50}{100} = \frac{1}{2}$
$33\frac{1}{3}$ out of 100 = $33\frac{1}{3}$%	$\frac{33.\overline{3}}{100} = \frac{1}{3}$
25 out of 100 = 25%	$\frac{25}{100} = \frac{1}{4}$
20 out of 100 = 20%	$\frac{20}{100} = \frac{1}{5}$
10 out of 100 = 10%	$\frac{10}{100} = \frac{1}{10}$
1 out of 100 = 1%	$\frac{1}{100} = \frac{1}{100}$
$66\frac{2}{3}$ out of 100 = $66\frac{2}{3}$%	$\frac{66.\overline{6}}{100} = \frac{2}{3}$
75 out of 100 = 75%	$\frac{75}{100} = \frac{3}{4}$

You can write fractions as percents and percents as fractions.

CONVERTING A FRACTION TO A PERCENT

Express $\frac{4}{5}$ as a percent.

- Set up a proportion.

  $\frac{4}{5} = \frac{n}{100}$

- Solve the proportion.

  $5n = 4 \times 100 \qquad 5n = 400 \qquad n = 80$

- Express as a percent.

  $\frac{80}{100} = 80\%$

  $\frac{4}{5} = 80\%$

Check It Out

Convert each fraction to a percent.

1. $\frac{3}{5}$ 2. $\frac{3}{10}$ 3. $\frac{9}{10}$ 4. $2\frac{2}{25}$

Changing Percents to Fractions

To change from a percent to a fraction, you write a fraction with the percent as the numerator and 100 as the denominator, then express the fraction in lowest terms.

CHANGING A PERCENT TO A FRACTION

Express 35% as a fraction.

- Change the percent directly to a fraction with a denominator of 100. The number of the percent becomes the numerator of the fraction.

$$35\% = \frac{35}{100}$$

- Simplify, if possible.

$$\frac{35}{100} = \frac{7}{20}$$

35% expressed as a fraction is $\frac{7}{20}$.

Check It Out

Change each percent to a fraction in lowest terms.

5. 17%   6. 5%   7. 36%   8. 64%

Changing Mixed Number Percents to Fractions

To change a mixed number percent to a fraction, first change the mixed number to an *improper fraction* (p. 107).

Express $25\frac{1}{2}\%$ as a fraction.

- Change the mixed number to an improper fraction.

$$25\frac{1}{2}\% = \frac{51}{2}\%$$

- Multiply the percent by $\frac{1}{100}$.

$$\frac{51}{2} \times \frac{1}{100} = \frac{51}{200}$$

- Simplify, if possible.

$$25\frac{1}{2}\% = \frac{51}{200}$$

Check It Out

Write the mixed number percent as a fraction.

9. $32\frac{1}{4}\%$ 10. $47\frac{3}{4}\%$ 11. $123\frac{1}{8}\%$

Percents and Decimals

Percents can be expressed as decimals and decimals can be expressed as percents. *Percent* means part of a hundred or hundredths.

CHANGING DECIMALS TO PERCENTS

Express 0.7 as a percent.
- Multiply the decimal by 100.

    $0.7 \times 100 = 70$
- Add the percent sign.

    $0.7 \rightarrow 70\%$

A Shortcut for Changing Decimals to Percents

Change 0.7 to a percent.
- Move the decimal point two places to the right. Add zeros, if necessary.

    $0.7 \rightarrow 70.$
- Add the percent sign.

    $70. \rightarrow 70\%$

Check It Out

Write each decimal as a percent.

12. 0.27   13. 0.007
14. 0.018   15. 1.5

Dollars, Pesos, Rupees, and Drachmas

Planning a trip outside the United States? Before you leave, you might want to change some of your U.S. money into money from each country you will be visiting. The rate of exchange varies from day to day and from country to country, depending on many factors.

The table lists some recent exchange rates for various places around the world. The number next to each country tells you what $1.00 in U.S. currency would equal when exchanged in the United States for foreign currency.

Country	Currency	Foreign Currency per U.S. dollar
Australia	Dollar	1.2763
England	Pound	0.6560
Japan	Yen	107.4
Kenya	Shilling	36.227
Mexico	Peso	7.53

Which countries' currencies are worth less than $1.00? If you purchase something for 15,000 yen, what is its equivalent price in U.S. dollars?
See Hot Solutions for answers.

2·9 RELATIONSHIPS

You can convert percents directly to decimals.

CHANGING PERCENTS TO DECIMALS

Change 4% to a decimal.
- Express the percent as a fraction with 100 as the denominator.

  $4\% = \frac{4}{100}$
- Change the fraction to a decimal by dividing the numerator by the denominator.

  $4 \div 100 = 0.04$

$4\% = 0.04$

A Shortcut for Changing Percents to Decimals
Change 7% to a decimal.
- Move the decimal point two places to the left.

  $7\% \longrightarrow .07.$
- Add zeros, if necessary.

  $7\% = 0.07$

Check It Out

Express each percent as a decimal.

16. 49% 17. 3% 18. 180% 19. .7%

Fractions and Decimals

Fractions can be written as either **terminating** or **repeating** decimals.

Fractions	Decimals	Terminating or Repeating
$\frac{1}{2}$	0.5	terminating
$\frac{1}{3}$	0.3333333...	repeating
$\frac{1}{6}$	0.166666...	repeating
$\frac{3}{5}$	0.6	terminating

CHANGING FRACTIONS TO DECIMALS

Write $\frac{4}{5}$ as a decimal.

- Divide the numerator of the fraction by the denominator.

    $4 \div 5 = 0.8$

The remainder is zero. The decimal 0.8 is a *terminating decimal.*

Write $\frac{1}{3}$ as a decimal.

- Divide the numerator of the fraction by the denominator.

    $1 \div 3 = 0.3333333...$

The decimal 0.3333333 is a *repeating decimal.*

- Place a bar over the digit that repeats.

    $0.\overline{3}$

Check It Out

Use a calculator to find a decimal for each fraction.

20. $\frac{3}{10}$ 　　　21. $\frac{7}{8}$ 　　　22. $\frac{1}{11}$

Changing Decimals to Fractions

Write a decimal as a fraction or mixed number.

- Write the decimal as a fraction with 100 as the denominator.

    $0.26 = \frac{26}{100}$

- Express the fraction in lowest terms.

    $\frac{26}{100} = \frac{13}{50}$

Check It Out

Write each decimal as a fraction.

23. 0.78 　　　24. 0.54 　　　25. 0.24

The Ups and Downs of Stocks

A corporation raises money by selling stocks—certificates that represent shares of ownership in the corporation. The stock page of a newspaper lists the high, low, and ending prices of the stock for the previous day. It also shows the overall fractional amount by which the price changed. A (+) sign indicates that the value of the stock increased; a (−) sign indicates the value decreased.

Suppose you see a listing on the stock page that shows the closing price of a stock was $21\frac{3}{4}$ with $+\frac{1}{4}$ next to it. What do those fractions mean? First, it tells you that the price of the stock was $21\frac{3}{4}$ dollars or $21.75. The $+\frac{1}{4}$ means the price went up $\frac{1}{4}$ of a dollar from the day before. Since $\frac{1}{4} \times \$1.00 = \$0.25$, the stock went up 25¢. To find the percent increase in the price of the stock, you have to first determine the original price of the stock. The stock went up $\frac{1}{4}$, so the original price is $21\frac{3}{4} - \frac{1}{4} = 21\frac{1}{2}$. What is the percent of increase to the nearest whole percent? See Hot Solutions for answer.

2·9 EXERCISES

Change each fraction to a percent.

1. $\frac{3}{12}$ 2. $\frac{13}{20}$ 3. $\frac{63}{100}$

4. $\frac{11}{50}$ 5. $\frac{7}{20}$

Change each percent to a fraction in lowest terms.

6. 24%   7. 62%   8. 33%

9. 10%   10. 85%

Write a decimal as a percent.

11. 0.6   12. 0.33   13. 0.121

14. 0.64   15. 2.5

Write each percent as a decimal.

16. 27%   17. 14.5%   18. 17%

19. 3%   20. 27.4%

Change each fraction to a decimal. Use a bar to show repeating decimals.

21. $\frac{5}{16}$ 22. $\frac{2}{3}$ 23. $\frac{3}{8}$

24. $\frac{2}{5}$ 25. $\frac{1}{25}$

Write each decimal as a fraction.

26. 0.76   27. 0.88   28. 0.9

29. 2.5   30. 0.24

31. One survey at Jefferson Middle School said that 45% of the seventh grade students wanted the spring dance to be a semiformal. Another survey reported that $\frac{9}{20}$ of the seventh grade students wanted the spring dance to be a semiformal. Could both surveys be correct? Explain.

32. Price Savers is advertising 40% off the $129 in-line skates. Bottom Dollar is advertising the same in-line skates at $\frac{1}{4}$ off the price of $129. Which is the better buy? How much would you save with the better buy?

What have you learned?

You can use the problems and the list of words that follow to see what you have learned in this chapter. You can find out more about a particular problem or word by referring to the boldfaced topic number (for example, **2•2**).

Problem Set

1. Which fraction is equivalent to $\frac{15}{20}$? **2•1**
   A. $\frac{3}{4}$ B. $\frac{6}{6}$ C. $\frac{4}{5}$ D. $\frac{5}{4}$
2. Which fraction is greater, $\frac{1}{13}$ or $\frac{3}{21}$? **2•2**
3. Write the improper fraction $\frac{17}{9}$ as a mixed number. **2•2**

Add or subtract. Write your answers in lowest terms. **2•3**

4. $\frac{4}{5} + \frac{2}{7}$ 5. $3\frac{1}{8} - \frac{1}{2}$ 6. $4 - 2\frac{1}{8}$

Multiply or divide. **2•3**

7. $\frac{3}{5} \times \frac{5}{8}$ 8. $\frac{5}{7} \div 3\frac{1}{2}$ 9. $2\frac{3}{4} \times \frac{5}{9}$

10. Give the place value of 3 in 251.034. **2•5**
11. Write 3.205 in expanded form. **2•5**
12. Write as a decimal: two hundred two and twenty-two thousandths. **2•5**
13. Write the following numbers in order from least to greatest: 0.880; 0.080; 0.088; 0.808 **2•5**

Find each answer. **2•6**

14. 11.44 + 2.834
15. 12.5 − 1.09
16. 8.07 × 5.6
17. 0.792 ÷ 0.22

Use a calculator. Round to the nearest tenth. **2•8**

18. What percent of 115 is 40?
19. Find 22% of 66.   20. 8 is 32% of what number?

21. Of the 12 girls on the basketball team, 8 play regularly. What percent play regularly? **2•8**
22. A soccer stadium has a seating capacity of 28,275 seats. 28% of the tickets are taken by season-ticket holders. How many seats are taken by season-ticket holders? **2•8**

Write each decimal as a percent. **2•9**
23. 0.85   24. 0.04

Write each fraction as a percent. **2•9**
25. $\frac{1}{100}$ 26. $\frac{160}{100}$

Write each percent as a decimal. **2•9**
27. 15%   28. 5%

Write each percent as a fraction in lowest terms. **2•9**
29. 80%   30. 8%

WRITE DEFINITIONS FOR THE FOLLOWING WORDS.

hot **words**

benchmark **2•7**
common
    denominator
    2•1
cross product **2•1**
denominator **2•1**
discount **2•8**
equivalent **2•1**

equivalent
    fractions **2•1**
estimate **2•6**
factor **2•4**
fraction **2•1**
greatest common
    factor **2•1**
improper fraction
    2•1
least common
    multiple **2•1**
mixed number **2•1**

numerator **2•1**
percent **2•7**
place value **2•5**
product **2•4**
ratio **2•7**
reciprocal **2•4**
repeating decimal
    2•9
terminating
    decimal **2•9**
whole numbers
    2•5

Powers and Roots

$2^3 = 8$

What do you already know?

You can use the problems and the list of words that follow to see what you already know about this chapter. The answers to the problems are in Hot Solutions at the back of the book, and the definitions of the words are in Hot Words at the front of the book. You can find out more about a particular problem or word by referring to the boldfaced topic number (for example, **3•2**).

Problem Set

Write each multiplication, using an exponent. **3•1**

1. $3 \times 3 \times 3 \times 3 \times 3$
2. $a \times a \times a$
3. $9 \times 9 \times 9$
4. $x \times x \times x \times x \times x \times x \times x \times x$

Evaluate each square. **3•1**

5. 3^2
6. 7^2
7. 4^2
8. 8^2

Evaluate each cube. **3•1**

9. 3^3
10. 4^3
11. 6^3
12. 9^3

Evaluate each power of 10. **3•1**

13. 10^4
14. 10^6
15. 10^{10}
16. 10^8

Evaluate each square root. **3•2**

17. $\sqrt{25}$
18. $\sqrt{64}$
19. $\sqrt{100}$
20. $\sqrt{81}$

Estimate each square root between two consecutive numbers. **3•2**
21. $\sqrt{31}$
22. $\sqrt{10}$
23. $\sqrt{73}$
24. $\sqrt{66}$

Estimate each square root to the nearest thousandth. **3•2**
25. $\sqrt{48}$
26. $\sqrt{57}$
27. $\sqrt{89}$
28. $\sqrt{98}$

Evaluate each cube root. **3•2**
29. $\sqrt[3]{27}$
30. $\sqrt[3]{125}$
31. $\sqrt[3]{216}$
32. $\sqrt[3]{1000}$

Write each number in scientific notation. **3•3**
33. 36,000,000
34. 600,000
35. 80,900,000,000
36. 540

Write each number in standard form. **3•3**
37. 5.7×10^6
38. 1.998×10^3
39. 7×10^8
40. 7.34×10^5

CHAPTER 3

hot **words**

	cube root **3•2**	scientific notation **3•3**
	exponent **3•1**	
area **3•1**	factor **3•1**	square **3•1**
base **3•1**	perfect square **3•2**	square root **3•2**
cube **3•1**	power **3•1**	volume **3•1**

WHAT DO YOU KNOW?

3·1 Powers and Exponents

Exponents

Multiplication, as you know, is the shortcut for showing a repeated addition: $4 \times 2 = 2 + 2 + 2 + 2$. A shortcut for showing the repeated multiplication $2 \times 2 \times 2 \times 2$ is to write 2^4. The 2 is the factor to be multiplied, called the **base.** The 4 is the **exponent,** which tells you how many times the base is to be multiplied. The expression can be read as "2 to the fourth **power.**" When you write an exponent, it is written slightly higher than the base and the size is usually a little smaller.

MULTIPLICATION USING EXPONENTS

Write the multiplication $8 \times 8 \times 8 \times 8 \times 8 \times 8$ using an exponent.

- Check that the same **factor** is being used in the multiplication.

    All the factors are 8.

- Count the number of times 8 is being multiplied.

    There are 6 factors of 8.

- Write the multiplication using an exponent.

    Since the factor 8 is being multiplied 6 times, write 8^6.

$8 \times 8 \times 8 \times 8 \times 8 \times 8 = 8^6$

Check It Out

Write each multiplication using an exponent.
1. $3 \times 3 \times 3 \times 3$
2. $7 \times 7 \times 7 \times 7$
3. $a \times a \times a \times a \times a \times a$
4. $z \times z \times z \times z \times z$

Evaluating the Square of a Number

The **square** of a number means to apply the exponent 2 to a base. The square of 3, then, is 3^2. To evaluate 3^2, identify 3 as the base and 2 as the exponent. Remember, the exponent tells you how many times to use the base as a factor. So 3^2 means to use 3 as a factor 2 times:

$$3^2 = 3 \times 3 = 9$$

The expression 3^2 can be read as "3 to the second power." It can also be read as "3 squared."

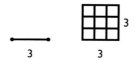

When a square is made from a segment whose length is 3, the **area** of the square is $3 \times 3 = 3^2 = 9$.

EVALUATING THE SQUARE OF A NUMBER

Evaluate 8^2.

- Identify the base and the exponent.

    The base is 8 and the exponent is 2.

- Write the expression as a multiplication.

    $$8^2 = 8 \times 8$$

- Evaluate.

    $$8 \times 8 = 64$$

    $$8^2 = 64$$

Check It Out

Evaluate each square.

5. 4^2 6. 8^2

7. 3 squared   8. 10 squared

Evaluating the Cube of a Number

To make the **cube** of a number means to apply the exponent 3 to a base. The cube of 2, then, is 2^3. Evaluating cubes is very similar to evaluating squares. For example, if you wanted to evaluate 2^3, notice that 2 is the base and 3 is the exponent. Remember, the exponent tells you how many times to use the base as a factor. So 2^3 means to use 2 as a factor 3 times:

$$2^3 = 2 \times 2 \times 2 = 8$$

The expression 2^3 can be read as "2 to the third power." It can also be read as "2 cubed."

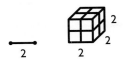

When a cube has edges of length 2, the **volume** of the cube is $2 \times 2 \times 2 = 2^3 = 8$.

EVALUATING THE CUBE OF A NUMBER

Evaluate 4^3.

- Identify the base and the exponent.

    The base is 4 and the exponent is 3.

- Write the expression as a multiplication.

    $$4^3 = 4 \times 4 \times 4$$

- Evaluate.

    $$4 \times 4 \times 4 = 64$$

$$4^3 = 64$$

Check It Out

Evaluate each cube.

9. 5^3

10. 10^3

11. 8 cubed

12. 6 cubed

Powers of Ten

Our decimal system is based on 10. For each factor of 10, the decimal point moves one place to the right.

$$2.11 \rightarrow 21.1 \qquad 19.05 \rightarrow 1,905 \qquad 7. \rightarrow 70$$
$$\times 10 \qquad\qquad\qquad \times 100 \qquad\qquad\qquad \times 10$$

When the decimal point is at the end of a number and the number is multiplied by 10, a zero is added at the end of the number.

Try to discover a pattern for the powers of 10.

Powers	As a Multiplication	Result	Number of Zeros
10^2	10×10	100	2
10^4	$10 \times 10 \times 10 \times 10$	10,000	4
10^5	$10 \times 10 \times 10 \times 10 \times 10$	100,000	5
10^8	$10 \times 10 \times 10 \times 10 \times 10 \times 10 \times 10 \times 10$	100,000,000	8

Notice that the number of zeros after the 1 is the same as the power of 10. This means that if you want to evaluate 10^7, you simply write a 1 followed by 7 zeros: 10,000,000.

Check It Out

Evaluate each power of 10.

13. 10^2 14. 10^6

15. 10^8 16. 10^3

3·1 POWERS AND EXPONENTS

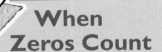

When Zeros Count

Usually you think that a zero means "nothing." But when zeros are related to a power of 10, you can get some fairly large numbers. A billion is the name for 1 followed by 9 zeros; a quintillion is the name for 1 followed by 18 zeros. You can write out all the zeros or use mathematical shorthand for these numbers.

$$1 \text{ billion} = 1{,}000{,}000{,}000 \text{ or } 10^9$$

$$1 \text{ quintillion} = 1{,}000{,}000{,}000{,}000{,}000{,}000 \text{ or } 10^{18}$$

What name would you use for 1 followed by one hundred zeros? According to the story, when the mathematician Edward Kasner asked his 9-year-old nephew to think up a name for this number, his nephew called it a *googol*. And that's the name used for 10^{100} today.

Suppose you could count at the rate of 1 number each second. If you started counting now and continued to count for 12 hours per day, it would take about 24 days to count to 1 million (1,000,000 or 10^6). Do you think you could count to a googol in your lifetime? See Hot Solutions for answer.

3·1 EXERCISES

Write each multiplication using an exponent.
1. $8 \times 8 \times 8$
2. $3 \times 3 \times 3 \times 3 \times 3 \times 3 \times 3$
3. $y \times y \times y \times y \times y \times y$
4. $n \times n \times n \times n \times n \times n \times n \times n \times n \times n$
5. 15×15

Evaluate each square.
6. 5^2
7. 14^2
8. 7^2
9. 1 squared
10. 20 squared

Evaluate each cube.
11. 5^3
12. 9^3
13. 11^3
14. 3 cubed
15. 8 cubed

Evaluate each power of 10.
16. 10^2
17. 10^{14}
18. 10^6

19. What is the area of a square whose sides have a length of 9?
    A. 18
    B. 36
    C. 81
    D. 729
20. What is the volume of a cube whose sides have a length of 5?
    A. 60
    B. 120
    C. 125
    D. 150

3·2 Square and Cube Roots

Square Roots

In mathematics, certain operations are opposites of each other. That is, one operation "undoes" the other. Addition undoes subtraction: $11 - 3 = 8$, so $8 + 3 = 11$. Multiplication undoes division: $16 \div 8 = 2$, so $2 \times 8 = 16$.

The opposite, or undoing, of squaring a number is finding the **square root.** You know that 3 squared $= 3^2 = 9$. The square root of 9 is the number that can be multiplied by itself to get 9, which is 3. The symbol for square root is $\sqrt{}$. Therefore $\sqrt{9} = 3$.

FINDING THE SQUARE ROOT

Find $\sqrt{25}$.

- Think, what number times itself makes 25?

   $5 \times 5 = 25$

- Find the square root.

   Since $5 \times 5 = 25$, the square root of 25 is 5.

Thus $\sqrt{25} = 5$.

Check It Out

Find each square root.

1. $\sqrt{9}$
2. $\sqrt{36}$
3. $\sqrt{81}$
4. $\sqrt{121}$

Estimating Square Roots

The table shows the first ten **perfect squares** and their square roots.

Perfect square	1	4	9	16	25	36	49	64	81	100
Square root	1	2	3	4	5	6	7	8	9	10

So then, how much is $\sqrt{50}$? In this problem, 50 is the square. In the table, 50 lies between 49 and 64. Thus $\sqrt{50}$ must be between $\sqrt{49}$ and $\sqrt{64}$, which would be between 7 and 8. To estimate the value of a square root, you can find the two consecutive numbers that the square root must be between.

ESTIMATING A SQUARE ROOT

Estimate $\sqrt{95}$.
- Identify the perfect squares that 95 is between.
  95 is between 81 and 100.
- Find the square roots of the perfect squares.
  $\sqrt{81} = 9$ and $\sqrt{100} = 10$.
- Estimate the square root.
  $\sqrt{95}$ is between 9 and 10.

Check It Out
Estimate each square root.
5. $\sqrt{42}$
6. $\sqrt{21}$
7. $\sqrt{5}$
8. $\sqrt{90}$

Better Estimates of Square Roots

If you want to know a better estimate for the value of a square root, you will want to use a calculator. Most calculators have a $\sqrt{}$ key for finding square roots.

On some calculators, the $\sqrt{}$ function is shown, not on a key, but above the x^2 key on the calculator's surface. If this is true for your calculator, you should also find a key that has either $\boxed{\text{INV}}$ or $\boxed{\text{2nd}}$ on it.

To use the $\sqrt{}$ function, you would press $\boxed{\text{INV}}$ or $\boxed{\text{2nd}}$ and then the key with $\sqrt{}$ above it.

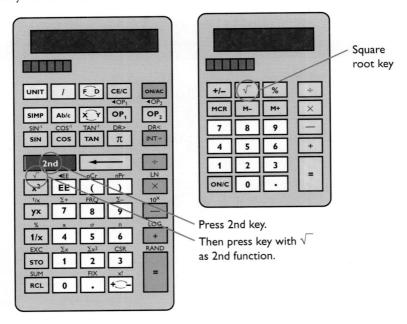

Square root key

Press 2nd key.
Then press key with $\sqrt{}$ as 2nd function.

When finding the square root of a number that is not a perfect square, the answer will be a decimal and the entire calculator display will be used. Generally you should round square roots to the nearest thousandth. Remember that the thousandths place is the third place after the decimal point.

See topics 9•1 and 9•2 for more about calculators.

3•2 SQUARE AND CUBE ROOTS

ESTIMATING THE SQUARE ROOT OF A NUMBER

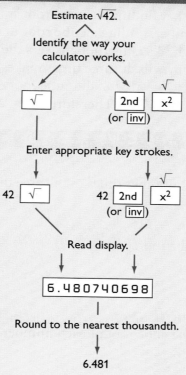

Estimate √42.

Identify the way your calculator works.

√   2nd √x²
           (or inv)

Enter appropriate key strokes.

42 √   42 2nd √x²
                  (or inv)

Read display.

6.480740698

Round to the nearest thousandth.

6.481

 Check It Out

Use a calculator to estimate each square root to the nearest thousandth.

9. $\sqrt{3}$

10. $\sqrt{47}$

11. $\sqrt{86}$

12. $\sqrt{97}$

Cube Roots

In the same way that finding a square root "undoes" the squaring of a number, finding a **cube root** undoes the cubing of a number. Finding a cube root answers the question, "What number times itself three times makes the cube?" Since 2 cubed = $2 \times 2 \times 2 = 2^3 = 8$, the cube root of 8 is 2. The symbol for cube root is $\sqrt[3]{}$. Therefore $\sqrt[3]{8} = 2$.

FINDING THE CUBE ROOT OF A NUMBER

Find $\sqrt[3]{216}$.

- Think, what number times itself three times will make 216?

  $6 \times 6 \times 6 = 216$

- Find the cube root.

  $\sqrt[3]{216} = 6$

Check It Out

Find the cube root of each number.

13. $\sqrt[3]{27}$
14. $\sqrt[3]{343}$
15. $\sqrt[3]{512}$
16. $\sqrt[3]{1000}$

3·2 EXERCISES

Find each square root.
1. $\sqrt{16}$
2. $\sqrt{49}$
3. $\sqrt{144}$
4. $\sqrt{25}$
5. $\sqrt{100}$

6. $\sqrt{31}$ is between which two numbers?
   A. 3 and 4
   B. 5 and 6
   C. 29 and 31
   D. None of these

7. $\sqrt{86}$ is between which two numbers?
   A. 4 and 5
   B. 8 and 9
   C. 9 and 10
   D. 83 and 85

8. $\sqrt{23}$ is between what two consecutive numbers?

9. $\sqrt{50}$ is between what two consecutive numbers?

10. $\sqrt{112}$ is between what two consecutive numbers?

Use a calculator to estimate each square root to the nearest thousandth.
11. $\sqrt{2}$
12. $\sqrt{20}$
13. $\sqrt{50}$
14. $\sqrt{75}$
15. $\sqrt{1000}$

Find the cube root of each number.
16. $\sqrt[3]{64}$
17. $\sqrt[3]{343}$
18. $\sqrt[3]{1728}$
19. $\sqrt[3]{1}$
20. $\sqrt[3]{27000}$

3·3 Scientific Notation

Using Scientific Notation

Often, in science and in mathematics, numbers are used that are very large. Large numbers often have many zeros at the end.

Large number: 450,000,000

Many zeros at the end

Instead of writing large numbers with all the zeros and possibly forgetting one of them, **scientific notation** was developed so that large numbers could be written without including all of the zeros. Scientific notation uses *powers of 10* (p. 169). To write a number in scientific notation, move the decimal point in the number so that only one digit is to the left of the decimal. Recall that each factor of 10 moves the decimal point one place to the right. Multiply the number by the correct power of 10.

WRITING A VERY LARGE NUMBER IN SCIENTIFIC NOTATION

Write 4,250,000,000 in scientific notation.

- Move the decimal point so that only one digit is to the left of the decimal.

    4.250000000.

- Count the number of decimal places that the decimal has to be moved to the left.

    4.250000000.
    9 places

- Write the number without the ending zeros, and multiply by the correct power of 10.

    4.25×10^9

Check It Out
Write each number in scientific notation.
1. 53,000
2. 4,000,000
3. 70,800,000,000
4. 26,340,000

Converting from Scientific Notation to Standard Form

Remember that each factor of 10 moves the decimal point one place to the right. When the last digit of the number is reached, there may still be some factors of 10 remaining. Add a zero at the end of the number for each remaining factor of 10.

CONVERTING TO A NUMBER IN STANDARD FORM

Write 7.035×10^6 in standard form.

• The exponent tells how many places to move the decimal point.

    The decimal point moves to the right 6 places.

• Move the decimal point the exponent number of places to the right. Add zeros at the end of the number if needed.

$$7.035000.$$
Move the decimal point
to the right 6 places.

• Write the number in standard form.
$$7.035 \times 10^6 = 7,035,000$$

Check It Out
Write each number in standard form.
5. 6.7×10^4
6. 2.89×10^8
7. 1.703×10^5
8. 8.52064×10^{12}

Bugs

Insects are the most successful form of life on Earth. About one million have been classified and named. It is estimated that there are up to four million more. That's not total insects we are talking about; that's different *kinds* of insects!

Insects have adapted to life in every imaginable place. They live in the soil, in the air, on the bodies of plants, animals, and other insects, on the edges of salt lakes, in pools of oil, and in the hot waters of hot springs.

Insects range in length from the stick-insect which can be over 15 inches long to a tiny parasitic fly less than .01 inch in length. Insects show an incredible range of adaptations. The tiny midge can beat its wings 62,000 times a minute. A flea can jump 130 times its height. Ants are so social they have been known to live in colonies with 1,000,000 queens and 300,000,000 workers.

Estimates are that there are 200,000,000 insects for each person on the planet. Given a world population of approximately 6,000,000,000, just how many insects do we share the earth with? Use a calculator to arrive at an estimate. Express the number in scientific notation. See Hot Solutions for answer.

3·3 SCIENTIFIC NOTATION

3•3 EXERCISES

Write each number in scientific notation.
1. 630,000
2. 408,000,000
3. 80,000,000
4. 15,020,000,000,000
5. 350
6. 7,060
7. 10,504,000
8. 29,000,100,000,000,000

Write each number in standard form.
9. 4.2×10^7
10. 5.71×10^4
11. 8.003×10^{10}
12. 5×10^8
13. 9.4×10^2
14. 7.050×10^3
15. 5.0203×10^9
16. 1.405×10^{14}

17. Which of the following expresses the number 5,030,000 in scientific notation?
    A. 5×10^6 B. 5.03×10^6
    C. 50.3×10^5 D. None of these
18. Which of the following expresses the number 3.09×10^7 in standard form?
    A. 30,000,000   B. 30,900,000
    C. 3,090,000,000   D. None of these
19. When written in scientific notation, which of the following numbers will have the greatest power of 10?
    A. 93,000   B. 408,000
    C. 5,556,000   D. 100,000,000
20. In scientific notation, what place value does 10^6 represent?
    A. hundred thousands   B. millions
    C. ten millions   D. hundred millions

What have you learned?

You can use the problems and the list of words that follow to see what you have learned in this chapter. You can find out more about a particular problem or word by referring to the boldfaced topic number (for example, **3•2**).

Problem Set

Write each multiplication, using an exponent. **3•1**

1. $20 \times 20 \times 20 \times 20 \times 20$
2. $k \times k \times k \times k \times k \times k \times k$
3. $4 \times 4 \times 4 \times 4 \times 4 \times 4$
4. $y \times y$

Evaluate each square. **3•1**

5. 4^2
6. 8^2
7. 13^2
8. 4^2

Evaluate each cube. **3•1**

9. 11^3
10. 12^3
11. 22^3
12. 8^3

Evaluate each power of 10. **3•1**

13. 10^3
14. 10^6
15. 10^8
16. 10^{12}

Evaluate each square root. **3•2**

17. $\sqrt{4}$
18. $\sqrt{36}$
19. $\sqrt{144}$
20. $\sqrt{100}$

Estimate each square root between two consecutive numbers. **3•2**

21. $\sqrt{55}$
22. $\sqrt{19}$
23. $\sqrt{99}$
24. $\sqrt{14}$

Use a calculator to estimate each square root to the nearest thousandth. **3•2**
25. $\sqrt{50}$
26. $\sqrt{18}$
27. $\sqrt{73}$
28. $\sqrt{7}$

Evaluate each cube root. **3•2**
29. $\sqrt[3]{64}$
30. $\sqrt[3]{216}$
31. $\sqrt[3]{10648}$
32. $\sqrt[3]{1331}$

Write each number in scientific notation. **3•3**
33. 2,902,000
34. 113,020
35. 40,100,000,000,000
36. 8,060

Write each number in standard form. **3•3**
37. 1.02×10^3
38. 2.701×10^7
39. 3.01×10^{12}
40. 6.1×10^2

hot **words**

WRITE DEFINITIONS FOR THE FOLLOWING WORDS.

	cube root **3•2**	scientific notation **3•3**
area **3•1**	exponent **3•1**	
base **3•1**	factor **3•1**	square **3•1**
cube **3•1**	perfect square **3•2**	square root **3•2**
	power **3•1**	volume **3•1**

WHAT HAVE YOU LEARNED?

hot topics 4

Data, Statistics, and Probability

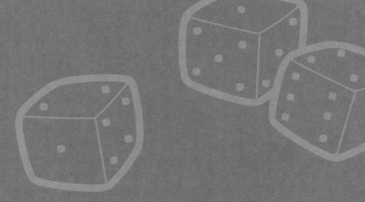

What do you already know?

You can use the problems and the list of words that follow to see what you already know about this chapter. The answers to the problems are in Hot Solutions at the back of the book, and the definitions of the words are in Hot Words at the front of the book. You can find out more about a particular problem or word by referring to the boldfaced topic number (for example, **4•2**).

Problem Set

1. Sonja walked through her neighborhood and stopped at her friends' houses to ask whether they planned to vote. Is this a random sample? **4•1**
2. Rebecca asked people in the mall the following question: What do you think of the ugly new city hall building? Was her question biased or unbiased? **4•2**

Use the following graph to answer items 3–5. **4•2**

Hulleah checked out several exercises and recorded the number of minutes she would have to do each one to burn 300 calories.

```
2 | 6 8 9
3 | 0 2 8 8 8 8 8 8 8
4 | 2 3 3 3 3
5 | 0 0 4
6 | 0 3 3
7 | 5
```

2 | 6 = 26 minutes

3. What kind of graph did Hulleah make?
4. What is the most common length of time she would have to exercise to burn 300 calories?
5. What exercise burns the most calories?

For items 6 and 7, use the following data about the number of minutes spent doing sit-ups. **4•2**

3 4 2 4 1 3 1 1 4 3 4 5 2 6 9 1

6. Make a frequency graph of the data.
7. Make a histogram of the data.

8. Hani had the following scores on his math tests: 90, 85, 88, 78, and 96. What is the range of scores? **4•4**
9. Find the mean and median of the scores in item 8. **4•4**

10. $P(5, 1) = ?$ **4•5**
11. $C(8, 2) = ?$ **4•5**
12. What is the value of 7!? **4•5**

Use the following information to answer items 13–15. **4•6**

A bag contains 20 marbles—8 red, 6 blue, 6 yellow.
13. One marble is drawn. What is the probability it is blue or yellow?
14. Two marbles are drawn. What is the probability they are both white?
15. A marble is drawn. Then a second one is drawn without replacement. What is the probability that both are blue?

CHAPTER 4

hot **words**

average **4•4**
bimodal distribution **4•3**
box plot **4•2**
circle graph **4•2**
combination **4•5**
correlation **4•3**
dependent events **4•6**
distribution **4•3**
double-bar graph **4•2**
event **4•6**
experimental probability **4•6**
factorial **4•5**
flat distribution **4•3**
histogram **4•2**
independent events **4•6**
leaf **4•2**
line graph **4•2**
mean **4•4**
median **4•4**
mode **4•4**

normal distribution **4•3**
outcome **4•6**
outcome grid **4•6**
percent **4•2**
permutation **4•5**
population **4•1**
probability **4•6**
probability line **4•6**
random sample **4•1**
range **4•4**
sample **4•1**
sampling with replacement **4•6**
scatter plot **4•3**
skewed distribution **4•3**
spinner **4•5**
stem **4•2**
stem-and-leaf plot **4•2**
strip graph **4•6**
survey **4•1**
table **4•1**
tally marks **4•1**
theoretical probability **4•6**
tree diagram **4•5**

4·1 Collecting Data

Surveys

Have you ever been asked to name your favorite movie? Or asked what kind of pizza you like? These kinds of questions are often asked in **surveys.** A statistician studies a group of people or objects, called a **population.** They usually get information from a small part of the population, called a **sample.**

In a survey, 200 seventh graders in Lakeville were chosen at random and asked what their favorite subject is in school. The following bar graph shows the percent of students who named each favorite subject.

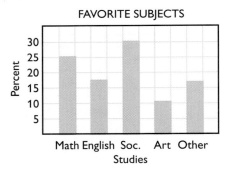

FAVORITE SUBJECTS

In this case, the population is all seventh graders in Lakeville. The sample is the 200 students who were actually asked to tell each favorite subject.

In any survey,
- The population consists of the people or objects about which information is desired.
- The sample consists of the people or objects in the population that are actually studied.

Check It Out

Identify the population and the size of the sample:
1. 500 adults who were registered to vote were surveyed at random to find out if they were in favor of increasing taxes to pay for education.
2. 200 fish in Sunshine Lake were captured, tagged, and released.

Random Samples

When you choose a sample to survey for data, be sure the sample is representative of the population. You also should be sure it is a **random sample**, where each person in the population has an equal chance of being included.

Ms. Landover wanted to find out whether her students wanted to go to a science museum, a fire station, the zoo, or a factory on a field trip. To pick a sample, she wrote the names of her students on cards and drew 20 cards from a bag. She then asked those 20 students where they wanted to go on the field trip.

FINDING A RANDOM SAMPLE

Determine whether Ms. Landover's sample is random.
*   Define the population.
    The population is Ms. Landover's class.
*   Define the sample.
    The sample consists of 20 students.
*   Determine if the sample is random.

Because every student in Ms. Landover's class had the same chance of being chosen, the sample is random.

Check It Out

3. Suppose you ask 20 people in a grocery store which grocery store is their favorite. Is the sample random?

4. A student assigns numbers to his 24 classmates and then writes the numbers on a slip of paper. He draws 10 numbers and asks those students what their favorite TV program is. Is the sample random?

5. How do you think you could select a random sample of your classmates?

Questionnaires

When you write questions for a survey, it is important to be sure that the questions are not biased. That is, the questions should not assume anything or influence the answers. The following two questionnaires are designed to find out what kind of books your classmates like and what they like to do in the summer. The first questionnaire uses biased questions. The second questionnaire uses questions that are not biased.

Survey 1
  A. What mystery novel is your favorite?
  B. What TV programs do you like to watch in the summer?

Survey 2
  A. What kind of books do you like to read?
  B. What do you like to do in the summer?

To develop a questionnaire,
• Decide what topic you want to ask about.
• Define a population and decide how to select a sample from that population.
• Develop questions that are not biased.

Check It Out
6. Why is **A** in Survey 1 biased?
7. Why is **B** in Survey 2 better than **B** in Survey 1?
8. Write a question that asks the same thing as the following question but is not biased: Are you an interesting person who reads a lot of books?

Compiling Data

Once Ms. Landover collected the data from her students about field trip preferences, she had to decide how to show the results. As she asked students their preference, she used **tally marks** to tally the answers in a table. The following **table** shows their answers.

Preferred Field Trip	Number Of Students
Science museum	⦀⦀ |||
Zoo	||||
Fire station	⦀⦀
Factory	|||

To make a table to compile data,
• List the categories or questions in the first column or row.
• Tally the responses in the second column or row.

Check It Out
9. How many students chose the zoo?
10. Which field trip was chosen by the fewest students?
11. If Ms. Landover uses the survey to choose a field trip to take, which one should she choose? Explain.

The WorldPOPClock

The U.S. Bureau of the Census estimates how many people are in the world each second on their WorldPOPClock. The estimate is based on projected births and deaths around the world.

At 2:00 A.M. EST on March 2, 1997, the WorldPOPClock estimate was 5,825,618,337. Using the table below, calculate the world population as of the date you read this page.

Time Unit	Projected Increase
Year	79,178,194
Month	6,598,183
Day	216,927
Hour	9,039
Minute	151
Second	2.5

You can check your answer on the WorldPOPClock on the Internet. Go to http://www.census.gov/ipc/www/popwnote.html then click the link to WorldPOPClock.

4·1 EXERCISES

1. Two hundred seventh graders in Carroll Middle School were asked to identify their favorite after-school activity. Identify the population and the sample. How big is the sample?
2. Salvador knocked on 25 doors in his neighborhood. He asked the residents who answered if they were in favor of the idea of closing the ice-skating rink in their city. Is the sample random?
3. To choose teachers to be surveyed, Rita obtained a list of the names of all the teachers and wrote each name on a slip of paper. She placed the slips of paper in a bag and drew 10 names. Is the sample random?

Are the following questions biased? Explain.

4. Are you happy about the lovely plants being planted in the school yard?
5. How many times each week do you eat a school lunch?

Write questions that ask the same thing as the following questions but are not biased.

6. Are you thoughtful about not riding your bike on the sidewalk?
7. Do you like the long walk to school, or do you prefer to take the bus or get a ride?

Mr. Kemmeries asked his students which type of food they wanted at a class party and tallied the following information.

Type Of Food	Number of Seventh Graders	Number of Eighth Graders
Chicken fingers	ℍℍ IIII	ℍℍ ℍℍ II
Bagels	ℍℍ I	III
Pizza	ℍℍ ℍℍ II	ℍℍ ℍℍ
Sandwiches	ℍℍ II	ℍℍ I
Turkey franks	III	ℍℍ I

8. Which type of food was most popular? How many students preferred that type?
9. Which type of food was preferred by 13 students?
10. How many students were surveyed?

4·2 Displaying Data

Interpret and Create a Table

You know that statisticians collect data about people or objects. One way to show the data is to use a table. Here are the number of letters in each word in the first sentence of *Little House in the Big Woods*.

4 4 1 4 5 5 3 1 6 4 5 2 3 3 5 2 9 2 1 6 4 5 4 2 4

MAKING A TABLE

Make a table to organize the data about letters in the words.

- Name the first column or row *what* you are counting.

    Label the first row *Number of Letters*.

- Tally the amounts for each category in the second column or row.

Number of Letters	1	2	3	4	5	6	7	8	9
Number of Words	///	/////	///	₩₩ ₩₩ //	₩₩ ₩₩	//			/

- Count the tallies and record the number in the second column.

Number of Letters	1	2	3	4	5	6	7	8	9
Number of Words	3	4	3	7	5	2	0	0	1

The most common number of letters in a word was 4. Three words have 1 letter.

Check It Out

1. There are two letter counts in the table which are not represented by any words. Which numbers of letters are they?

2. Make a table with the following data to show the number of hours spent watching TV each day by a group of seventh graders.
    9 9 3 7 6 3 3 2 4 0 1 1 0 2 1 2 1 1 4 0 1 0

Interpret a Box Plot

A **box plot** shows data using the middle value of the data and the quartiles, or 25% divisions of the data. The following box plot shows stereo prices at Big Discount stereo store.

Stereo Prices at Big Discount

$50 $75 $100 $125 $150 $175 $200 $225

1st quartile

2nd quartile
(middle score)

3rd quartile

On a box plot, 50% of the prices are above the middle price and 50% are below it. The first-quartile price is the middle price of the bottom half of the prices. The third-quartile price is the middle price in the top half of the prices.

Using the boxplot, you can tell:
• The most expensive stereo is $200.
  The least expensive stereo is $50.
• The middle price is $125. The first-quartile price is $70, and the third-quartile price is $150.
• 50% of the prices are between $70 and $150.

 Check It Out

Use the following box plot to answer these questions.

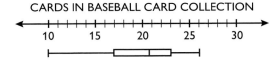

CARDS IN BASEBALL CARD COLLECTION

10   15   20   25   30

3. What is the greatest number of cards in a collection?
4. What percent of the collections have fewer than 17 cards?
5. What percent of the collections contain between 17 and 23 cards?

Interpret and Create a Circle Graph

Another way to show data is to use a **circle graph.** Aisha conducted a survey of her classmates. She found out that 24% walk to school, 32% take the school bus, 20% get a ride, and 24% ride their bikes. She wanted to make a circle graph to show her data.

MAKING A CIRCLE GRAPH

To make a circle graph,

- Find what **percent** of the whole each part of the data is.

    In this case, the percents are given.

- Multiply each percent by 360°, the number of degrees in a circle.

    $360° \times 24\% = 86.4°$ $\qquad$ $360° \times 20\% = 72°$

    $360° \times 32\% = 115.2°$

- Draw a circle, measure each central angle, and complete the graph.

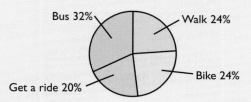

HOW WE GET TO SCHOOL

Bus 32% — Walk 24%

Get a ride 20% — Bike 24%

From the graph, you can see that more than half of the students travel by bus or get a ride.

Check It Out

Use the circle graph to answer items 6 and 7.

6. About what fraction of the pets were dogs?

7. About what fraction of the pets were not cats or dogs?

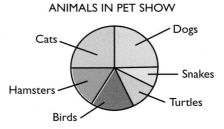

ANIMALS IN PET SHOW

Cats
Dogs
Snakes
Hamsters
Turtles
Birds

8. The following entries were received at the county fair. Make a circle graph to show the entries.

Pies: 76　　Jams: 36　　Cakes: 50　　Bread: 38

And the Winner Is...

At the end of February 1996, a performer won three Grammys.

Album Sales

75,000
50,000
25,000
0

4　11　18　25　3　10　17　24　31
February　March

Based on the graph, how did winning the Grammy awards affect album sales? How often were the sales tabulated? What kind of graph is this? See Hot Solutions for answers.

4·2 DISPLAYING DATA

Interpret and Create a Frequency Graph

You have used tally marks to show data. Suppose you collect the following information about the number of people in your classmates' families.

4 2 3 6 5 6 3 2 4 3 5 5 3 7 5 3 4 3

To make a frequency graph, you can place X's above a number line.

• Draw a number line showing the numbers in your data set.

> You would draw a number line showing the numbers 2 through 7.

• Place an X to represent each result above the appropriate number on the number line.

> For this frequency graph, you will put an X above the number in each family.

• Title the graph.

> You can call it "Number in Our Families."

Your frequency graph should look like this:

NUMBER IN OUR FAMILIES

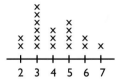

You can tell from your frequency graph that the most frequent number of members in a family is 3.

 Check It Out

9. What is the most people in a classmate's family?
10. How many classmates have 5 people in their family?
11. Make a frequency graph to show the number of letters in each word in the first sentence of *Little House in the Big Woods* (p. 194).

Interpret a Line Graph

You know that a *line graph* can be used to show changes in data over time. The following line graph shows the average number of wood ducks to visit a duck pond over the year.

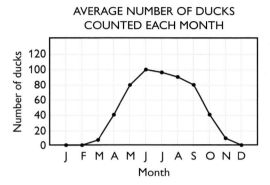

AVERAGE NUMBER OF DUCKS
COUNTED EACH MONTH

From the graph, you can see that there are fewer wood ducks in the area in winter months.

Check It Out

12. What was the greatest average number of ducks?
13. What month had the greatest average number of ducks?
14. Is this statement true or false: There were never more than 100 ducks in the pond.

Interpret a Stem-and-Leaf Plot

The following numbers show the number of wins by teams in March one year.

38 23 36 22 30 24 31 26 27 32 27 35 23 40 23 32 25 31
28 28 26 31 28 41 25 29

It is hard to tell much about the wins when they are displayed like this. You know that you could make a table, a box chart, or a line graph to show this information. Another way to show the information is to make a **stem-and-leaf plot.** The following stem-and-leaf plot shows the number of wins.

```
2 | 2 3 3 3 4 5 5 6 6 7 7 8 8 8 9
3 | 0 1 1 1 2 2 5 6 8
4 | 0 1
```

2 | 2 means 22 wins

Notice that the tens digits appear in the left-hand columns. These are called **stems.** Each digit on the right is called a **leaf.** From looking at the plot, you can tell that most of the teams have from 20 to 29 wins.

Check It Out

Use the stem-and-leaf plot showing the number of books read by students in the summer reading contest at the library.

15. How many students participated in the contest?
16. How many books did the students read in all?
17. Two students read the same number of books. How many books was that?

```
1 | 0 2 3
2 | 0 2 2 4 5 8
3 |
4 | 1 3 4
```

4 | 1 means 41 books

Interpret and Create a Bar Graph

Another type of graph you can use to show data is called a *bar graph*. In this graph, either horizontal or vertical bars are used to show data. Consider the data showing the tallest buildings in the United States.

Building 1	346 m
Building 2	381 m
Building 3	343 m
Building 4	443 m
Building 5	419 m

You can make a bar graph to show these heights:

- Choose a vertical scale and decide what to place along the horizontal scale.

    In this case, the vertical scale can show meters in increments of 50 m and the horizontal scale can show the buildings.
- Above each building, draw a bar of the appropriate height.
- Title the graph.

    You can call it "Heights of U.S. Skyscrapers."
    Your bar graph should look like this.

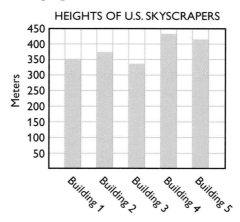

From the graph, you can see that the tallest building is Building 4.

Check It Out

18. What is the shortest building shown?

19. Using the following data about the number of middle-school students at Randall School, make a bar graph.

6th Grade: 230   7th Grade: 182   8th Grade: 199

Interpret a Double-Bar Graph

You know that you can show information in a bar graph. If you want to show information about two or more things, you can use a **double-bar graph.** This graph shows how many students were on the honor roll at the end of last year and this year.

NUMBER OF STUDENTS ON HONOR ROLL

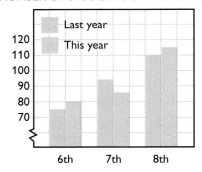

You can see from the graph that more eighth graders were on the honor roll this year than last year.

Check It Out

20. About how many students were on the honor roll at the end of this year?

21. Write a conclusion you can form from this graph.

Interpret and Create a Histogram

A special kind of bar graph that shows frequency of data is called a **histogram.** Suppose you asked several classmates how many books they have checked out from the school library and collected the following numbers:
3 3 3 1 1 0 4 2 1 3 4 2 1 0 1 6

MAKING A HISTOGRAM

Create a histogram.

• Make a table showing frequencies.

Books	Tally	Frequency
0	//	2
1	/////	5
2	//	2
3	////	4
4	//	2
5		0
6	/	1

• Make a bar graph showing the frequencies.
• Title the graph.

You can call it "Books Checked Out from the Library."

Your histogram will look like this:

You can see from the histogram that no students have five books checked out.

Check It Out

22. How many students did you survey?
23. Make a histogram from the data about *Little House in the Big Woods* (p. 194).

Graphic Impressions

Humans can live longer than 100 years. But not all animals can live that long. A mouse, for example, has a maximum life span of three years, while a toad may live for 36 years.

Both these graphs compare the maximum life span of guppies, giant spiders, and crocodiles.

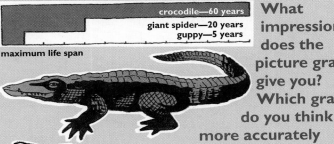

crocodile—60 years
giant spider—20 years
guppy—5 years

maximum life span

What impressions does the picture graph give you? Which graph do you think more accurately portrays the relative differences in the maximum life spans of these three animals? See Hot Solutions for answers.

4·2 EXERCISES

1. Make a table and a histogram to show the following data.
   Number of Times Students Purchased Hot Lunch
   4 3 5 1 4 0 3 0 2 3 1 5 4 3 1 5 0 2 3 1

2. How many students were surveyed?
3. Make a frequency graph to show the data in item 1.

4. Use your frequency graph to describe the number of hot lunches purchased.
5. Ryan spends $10 of his allowance each week on school lunch and $7 on CDs and magazines. He saves $3. Make a circle graph to show how Ryan spends his allowance and write a sentence about the graph.
6. The following stem-and-leaf plot shows the number of deer counted each day at a wildlife feeding station over a three-week period.

```
1 | 0 2 3 4 4 4 6 8
2 | 0 0 3 4 4 4 5 6 8 8
3 | 1 1 2
```

1|3 means 13

Draw a conclusion from the plot.

7. Six seventh graders ran laps around the school. Vanessa ran 8 laps, Tomas ran 3, Vedica ran 4, Samuel ran 6, Forest ran 7, and Tanya ran 2. Make a bar graph to show this information.
8. The box plot shows math test scores. What is the middle score? 50% of the scores are between 43 and what score?

MATH TEST SCORES

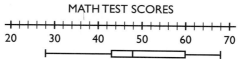

4•3 Analyzing Data

You can plot points on a *coordinate graph* (p. 300) to graph data. The graph can show if the data are related to each other.

Correlation

Suppose you collect some data in regard to 10 students in your class. You can write the information you have collected as ordered pairs and then graph the ordered pairs. The following graphs show the data you collected:

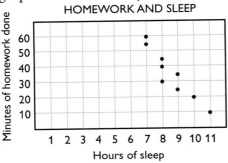

The graph (left) shows a negative **correlation.** The longer students spent on homework the less sleep they got. A downward trend in points shows a negative correlation.

The graph (right) shows a positive correlation. The longer students spent on homework, the higher their test scores were. An upward trend in points shows a positive correlation.

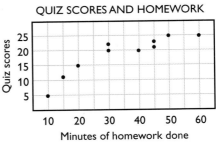

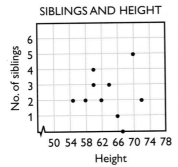

The graph (left) shows no correlation. The number of siblings had no relationship to student height. If there is no trend in graph points, there is no correlation.

Check It Out

1. Which of the following **scatter plots** shows a negative correlation?

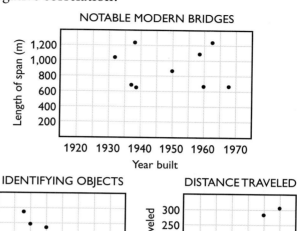

NOTABLE MODERN BRIDGES

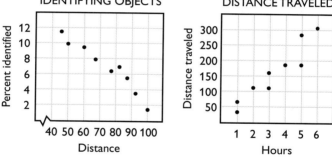

2. Describe the correlation in the scatter plot showing the relationship between the year a bridge was built and its length.
3. Which scatter plot shows a positive correlation?

Distribution of Data

Sometimes the histogram you draw is symmetrical. If you draw a curve over the histogram, the curve illustrates a **normal distribution.**

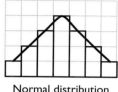
Normal distribution

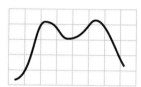
Skewed left

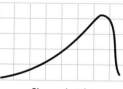
Skewed right

Data is not always evenly distributed. Often a histogram has a **skewed distribution.** Two of the histograms above show **distributions** that are skewed to the left and to the right. The histograms below show other distribution variations.

Bimodal distribution

Flat distribution

Check It Out

Identify each type of distribution as normal, skewed to the right, skewed to the left, bimodal, or flat.

4.
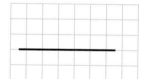

5.
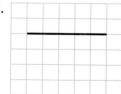

6.

7.
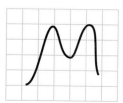

8.

4•3 EXERCISES

1. Choose the best answer.

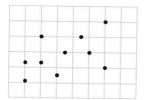

   A. Positive correlation
   B. Negative correlation
   C. No correlation

Describe the correlation in each of the following scatter plots.

2.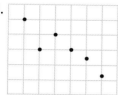
3.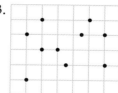
4.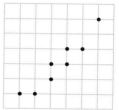

Tell whether each of the following distributions is normal, skewed to the right, skewed to the left, bimodal, or flat.

5.

6.

7.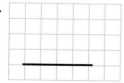

8.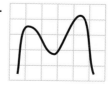

4·4 **Statistics**

Cleve collected the following data about the amounts of money his friends raised selling cookies.

$10, $10, $12, $15, $18, $19, $21, $23, $25, $28, $32, $35, $38, $39, $140

Cleve said his classmates typically raised $10, but Kimiko disagreed. She said the typical amount was $23. A third classmate, Gabe, said they were both wrong—the typical amount was $31. Each was correct, because each was using a different common measure.

Mean

One measure of data is the **mean.** To find the mean, or **average,** add the amounts the students raised and divide by the number of amounts.

FINDING THE MEAN

Find the mean amount raised by each friend selling cookies.

- Add the amounts.

  $10 + $10 + $12 + $15 + $18 + $19 + $21 + $23 + $25 + $28 + $32 + $35 + $38 + $39 + $140 = $465

- Divide the total by the number of amounts.

  In this case, there are 15 amounts:

  $465 ÷ 15 = $31

The mean amount raised was $31. Gabe used the mean when he said the typical amount raised was $31.

Check It Out
Find the mean:
1. 14, 12.5, 11.5, 4, 8, 14, 14, 10
2. 45, 100, 82, 83, 100, 46, 90
3. 321, 354, 341, 337, 405, 399
4. Winnie priced jeans at five stores. They were priced at $40, $80, $51, $64, and $25. Find the mean price.

Median

You see that you can find the mean by adding all the amounts and dividing by the number of amounts. Another way to look at numbers is to find the median. The **median** is the middle number in the data when the numbers are arranged in order. Let's look again at the amounts raised by the friends selling cookies.

$10, $10, $12, $15, $18, $19, $21, $23, $25, $28, $32, $35, $38, $39, $140

FINDING THE MEDIAN

Find the median amount made by the friends selling cookies.

• Arrange the data in numerical order from least to greatest or greatest to least.

Looking at the amounts raised, we can see they are already arranged in order.

• Find the middle number.

There are 15 numbers. The middle number is $23 because there are 7 numbers above $23 and 7 below it.

Kimiko was using the median when she described the typical amount raised.

When the number of amounts is even, you can find the median by finding the mean of the two middle numbers. So to find the median of the numbers 2, 6, 4, 3, 5, and 9, you must find the two numbers in the middle.

FINDING THE MEDIAN OF AN EVEN NUMBER OF DATA

Find the median of 2, 6, 4, 3, 5, and 9.

- Arrange the numbers in order from least to greatest or greatest to least.

    2, 3, 4, 5, 6, 9   or   9, 6, 5, 4, 3, 2

- Find the mean of the two middle numbers.

    The two middle numbers are 4 and 5:

    $(4 + 5) \div 2 = 4.5$

The median is 4.5. Half the numbers are greater than 4.5 and half the numbers are less than 4.5.

Check It Out

Find the median:

5. 11, 18, 11, 5, 17, 18, 8
6. 5, 9, 2, 6
7. 45, 48, 34, 92, 88, 43, 58
8. A naturalist measured the heights of 10 sequoia trees and got the following results, in feet: 260, 255, 275, 241, 238, 255, 221, 270, 265, and 250. Find the median height.

Mode

To describe a set of numbers, you can use the mean or use the median, which is the middle number. Another way to describe a set of numbers is to use the mode. The **mode** is the number in the set that occurs most often. Let's look again at the amounts raised by the friends selling cookies:

> $10, $10, $12, $15, $18, $19, $21, $23, $25, $28, $32, $35, $38, $39, $140

To find the mode, look for the number that appears most frequently.

FINDING THE MODE

Find the mode of the amounts raised by the friends selling cookies.

- Arrange the numbers in order or make a frequency table of the numbers.

   The numbers are arranged in order above.

- Select the number that appears most frequently.

   The most frequent amount raised is $10.

So Gabe was using the mode when he described the typical amount his friends raised selling cookies.

A group of numbers may have no mode or more than one mode. Data that have two modes is called *bimodal*.

Check It Out

Find the mode:

9. 21, 23, 23, 29, 27, 22, 27, 27, 24, 24
10. 3.8, 4.2, 4.2, 4.7, 4.2, 8.1, 1.5, 6.4, 6.4
11. 3, 9, 11, 11, 9, 3, 11, 4, 8
12. Soda in eight vending machines was selling for $0.75, $1, $1.50, $0.75, $0.95, $1, $0.75, and $1.25.

Range

Another measure used with numbers is the range. The **range** tells how far apart the greatest and least numbers in a set are. Consider the highest points on the seven continents:

Continent	Highest Elevation
Africa	19,340 ft
Antarctica	16,864 ft
Asia	29,028 ft
Australia	7,310 ft
Europe	18,510 ft
No. America	20,320 ft
So. America	22,834 ft

To find the range, you must subtract the least altitude from the greatest.

FINDING THE RANGE

Find the range of the highest elevations on the seven continents.

- Find the greatest and least values.

    The greatest value is 29,028 and the least value is 7,310.

- Subtract.

    $29,028 - 7,310 = 21,718$

The range is 21,718 ft.

Check It Out

Find the range:

13. 250, 300, 925, 500, 15, 600
14. 3.2, 2.8, 6.1, 0.4
15. 48°, 39°, 14°, 26°, 45°, 80°
16. The following number of students stayed for after-school activities one week: 28, 32, 33, 21, 18.

4·4 EXERCISES

Find the mean, median, mode, and range. Round to the nearest tenth.
  1. 3, 3, 4, 6, 6, 7, 8, 10, 10, 10
  2. 20, 20, 20, 20, 20, 20, 20
  3. 12, 9, 8, 15, 15, 13, 15, 12, 12, 10
  4. 76, 84, 88, 84, 86, 80, 92, 88, 84, 80, 78, 90
  5. Are any of the sets of data in items 1–4 bimodal? Explain.

  6. The highest point in Arizona is Mt. Humphreys, at 12,633 ft, and the lowest point, 70 ft, is on the Colorado River. What is the range in elevations?

  7. Moises had scores of 83, 76, 92, 76, and 93 on his history tests. Which of the mean, median, or mode do you think he should use to describe the test scores?

  8. Does the mean have to be a member of the set of data?

  9. The following numbers represent winning scores in baseball by the Treetown Tigers:
      3 2 4 2 30 2 1 2 4 7 2 1
     Find the mean, median, and mode of the scores. Round to the nearest tenth. Which measure best represents the data? Explain.

  10. Are you using the mean, median, or mode when you say that most of the runners finished the race in 5 minutes?

4·5 Combinations and Permutations

Tree Diagrams

You often need to be able to count outcomes. For example, suppose you have two **spinners**—one with the numbers 1 and 2 and another that is half-red and half-green. You want to find out all the possible outcomes if you spin the spinners. You can make a **tree diagram.**

To make a tree diagram, you list what can happen with the first spinner.

First
spinner

1
2

Then you list what can happen with the second spinner.

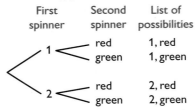

First spinner	Second spinner	List of possibilities
1	red	1, red
	green	1, green
2	red	2, red
	green	2, green

After listing the possibilities, you can count to see how many there are. In this case, there are four possible results.

To find the number of results or possibilities, you multiply the number of choices at each step:

$2 \times 2 = 4$

Check It Out
Draw a tree diagram. Check by multiplying.

1. Salvador has red, blue, and green shirts, and black, white, and gray slacks. How many possible outfits does he have?
2. Balls come in red, blue, and orange, and large and small. If a store owner has a bin for each type of ball, how many bins are there?
3. If you toss three coins, how many possible ways can they land?
4. How many possible ways can Jamie walk from school to home if she goes to the library on the way home?

School   Library   Home

Monograms

What are your initials? Do you have anything with your monogram on it? A *monogram* is a design that is made up of one or more letters, usually the initials of a name. Monograms often appear on stationary, towels, shirts, or jewelry.

How many different three-letter monograms can you make with the letters of the alphabet? Use a calculator to compute the total number. Don't forget to allow for repeat letters in the combination. See Hot Solutions for answer.

Permutations

The tree diagram shows ways in which things can be arranged, or listed. A listing in which the order is important is called a **permutation.** Suppose you want to take a picture of the dogs Spot, Brownie, and Topsy. They could be arranged in any of several ways. You can use a tree diagram to show the ways.

Dog on left	Dog in center	Dog on right	Arrangement list
Spot	Brownie — Topsy		SBT
	Topsy —— Brownie		STB
Brownie	Spot —— Topsy		BST
	Topsy —— Spot		BTS
Topsy	Spot —— Brownie		TSB
	Brownie — Spot		TBS

There are 3 choices for the first dog, 2 choices for the second, and 1 choice for the third, so the total number of permutations is $3 \times 2 \times 1 = 6$. Remember that Spot, Brownie, Topsy is a different permutation from Topsy, Brownie, Spot.

$P(3, 3)$ represents the number of permutations of 3 things taken 3 at a time. Thus $P(3, 3) = 6$.

FINDING PERMUTATIONS

Find $P(5, 3)$.

- Determine how many choices there are for each place.

  There are 5 choices for the first place, 4 for the second, and 3 for the third.

- Find the product.

  $5 \times 4 \times 3 = 60$

So $P(5, 3) = 120$.

Factorial Notation

You saw that to find the number of permutations of 3 things, you found the product $3 \times 2 \times 1$. The product $3 \times 2 \times 1$ is called 3 **factorial.** The shorthand notation for factorial is an exclamation point. So $3! = 3 \times 2 \times 1$.

Check It Out

Find each value.

5. $P(4, 3)$
6. $P(6, 6)$
7. The dog show has four finalists. In how many ways can four prizes be awarded?
8. One person from the 10-member swim team will represent the team at the city finals and another member will represent the team at the state finals. In how many ways can you choose the two people?

Find each value. Use a calculator if needed.

9. $8!$
10. $6! \div 4!$

Combinations

When you choose two delegates from a class of 35 to attend a science fair, the order is not important. That is, choosing Norma and Miguel is the same as choosing Miguel and Norma when you are choosing two delegates.

You can use the number of permutations to find the number of **combinations.** Say you and four friends (Jonah, Elena, Itay, and Kei) are going to the amusement park. There is room in your car for three of them. What different combinations of friends can ride with you?

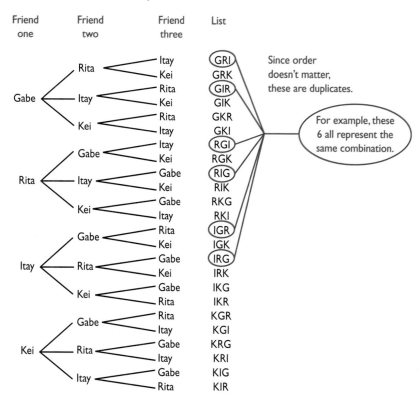

You have four choices for the first person, three choices for the second, and two choices for the third ($4 \times 3 \times 2 = 24$). But the order doesn't matter, so some combinations were counted too often! You need to divide by the number of different ways three objects can be arranged (3!).

$$\frac{4 \times 3 \times 2}{3 \times 2 \times 1} = 4$$

So there are four different groups of friends that can go in your car.

$C(4, 3)$ represents the number of combinations of four objects taken three at a time. Therefore $C(4, 3) = 4$.

FINDING COMBINATIONS

Find $C(5, 2)$.

• Find the permutations of the number of choices taken a number of times.

$P(5, 2)$

$P(5, 2) = 5 \times 4 = 20$

• Divide the permutations by the number of ways they can be arranged.

Two items can be arranged 2! ways.

$20 \div 2! = 20 \div 2 = 10$

So $C(5, 2) = 10$.

Check It Out

Find each value.

11. $C(8, 6)$

12. $C(12, 3)$

13. How many different combinations of four books can you choose from nine books?

14. Are there more combinations or permutations of three cats from a total of 12? Explain.

Lottery Fever

You read the headline. You say to yourself, "Somebody's *bound* to win this time." But the truth is, you would be wrong! The chances of winning a Pick-6 lottery are always the same, and very, very, very small.

Start with the numbers from 1 to 7. There are always 7 different ways to choose 6 out of 7 things. (Try it for yourself.) So, your chances of winning a 6-out-of-7 lottery would be $\frac{1}{7}$ or about 14.3%. Suppose you try using 6 out of 10 numbers. There are 210 different ways you can do that, making the likelihood of winning a 6-out-of-10 lottery $\frac{1}{210}$ or 0.4%. For a 6-out-of-20 lottery, there are 38,760 possible ways to pick 6 numbers, and only 1 of these would be the winner. That's about a 0.003% chance of winning. Get the picture?

The chances of winning a 6-out-of-50 lottery are 1 in 15,890,700 or 1 in about 16 million. For comparison, think about the chances that you will get struck by lightning—a rare occurrence. It is estimated that in the United States roughly 260 people are struck by lightning each year. Suppose the U.S. population is about 260 million. Would you be more likely to win the lottery or be struck by lightning? See Hot Solutions for answer.

4·5 EXERCISES

1. Make a tree diagram to show the results when you roll a number cube containing the numbers 1 through 6 and spin a spinner containing 1 and 2.

Find each value.

2. $P(6, 4)$
3. $C(10, 10)$
4. $P(8, 5)$
5. $C(8, 2)$
6. $6! \div 4!$
7. $P(7, 7)$

Solve.

8. Six friends want to play enough games of checkers to be sure every one plays everyone else. How many games will they have to play?

9. Eight people swim in the 200 m finals at an Olympic trial. Medals are given for first, second, and third place. How many ways are there to give the medals?

10. Determine if the following is a permutation or a combination.
   A. choosing first, second, and third place at an oratory contest among 15 people
   B. choosing four delegates from a class of 20 to attend Government Day

4·6 Probability

If you and a friend want to decide who goes first in a game, you might flip a coin. You and your friend have an equal chance of winning the toss. The **probability** of an event is a number from 0 to 1 that measures the chance that an event will occur.

Experimental Probability

The probability of an event is a number from 0 to 1. One way to find the probability of an event is to conduct an experiment. Suppose you want to know the probability of your winning a game of tennis. You play 12 games and win 8 of them. To find the probability of winning, you can compare the number of games you win to the number of games you play. In this case, the **experimental probability** that you will win is $\frac{8}{12}$, or $\frac{2}{3}$.

DETERMINING EXPERIMENTAL PROBABILITY

Find the experimental probability of finding your friend at the library when you go.

- Conduct an experiment. Record the number of trials and the result of each trial.

   Go to the library 10 different times and see if your friend is there. Suppose your friend is there 4 times.

- Compare the number of occurrences of one set of results to the number of trials. That is the probability for that set of results.

   Compare the number of times your friend was there to the total number of times you went to the library.

The experimental probability of finding your friend at the library for this test is $\frac{4}{10}$, or $\frac{2}{5}$.

Check It Out

A marble is drawn from a bag of 20 marbles. Each time, the marble was returned before the next one was drawn. The results are shown on the circle graph.

MARBLES DRAWN

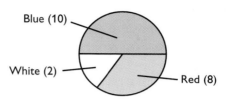

Blue (10)

White (2)

Red (8)

1. Find the experimental probability of getting a red marble.
2. Find the experimental probability of getting a white marble.
3. Place 50 counters of different colors in a bag. Find the experimental probability that you will choose a counter of one color. Compare your answers with others' answers.

Theoretical Probability

You know that when you want to find the experimental probability of tossing a head when you toss a coin, you can do the experiment and record the results. You can also find the **theoretical probability** if you consider the outcomes of the experiment. The **outcome** of an experiment is a result. The outcomes of tossing a coin are a "head" and a "tail." An **event** is a specific outcome, such as a head. So the probability of getting a head is:

$$\frac{\text{number of ways an event occurs}}{\text{number of outcomes}} = \frac{1}{2}$$

PROBABILITY

4•6

DETERMINING THEORETICAL PROBABILITY

Find the probability of rolling an even number when you roll a number cube containing the numbers 1–6.

- Determine the number of ways the event occurs.

    In this case, the event is getting an even number. There are three even numbers—2, 4, and 6—on the number cube.

- Determine the total number of outcomes. Use a list, multiply, or make a *tree diagram* (p. 216).

    There are six numbers on the cube.

- Use the formula:

    $P(\text{event}) = \dfrac{\text{Number of ways an event occurs}}{\text{Number of outcomes}}$

- Find the probability of the target event.

    Find the probability of rolling an even number, represented by $P(\text{even})$.

    $P(\text{even}) = \frac{3}{6} = \frac{1}{2}$

The probability of rolling an even number is $\frac{1}{2}$.

Check It Out

Find each probability. Use the spinner for items 5–6.

4. $P(\text{odd number})$ when tossing a 1–6 number cube
5. $P(\text{red})$ 6. $P(\text{green})$
7. The letters of the word *mathematics* are written on identical slips of paper and placed in a bag. If you draw a slip at random, what is the probability that it will be an *m*?

Expressing Probabilities

You can express a probability as a fraction, as shown before. But just as you can write a fraction as a decimal, ratio, or percent, you can also write a probability in any of those forms.

The probability of getting a head when you toss a coin is $\frac{1}{2}$. You can also express the probability as follows:

Fraction	Decimal	Ratio	Percent
$\frac{1}{2}$	0.5	1:2	50%

Check It Out

Express each of the following probabilities as a fraction, decimal, ratio, and percent.

8. the probability of drawing a vowel from cards containing the letters of the name *Washington*

9. the probability of getting a red marble when you are drawing a marble from a bag containing two red marbles and six blue ones

Strip Graphs

When you conduct an experiment, such as tossing a coin, you need to find a way to show the outcome of each toss. One way to show the outcomes of an experiment is to use a **strip graph**.

The following strip graph shows the result of tossing a coin.

| H | H | T | H | T | H | T | T |

The strip graph shows that the following results occurred on the first eight tosses: head, head, tail, head, tail, head, tail, and tail.

To make a strip graph,
• Draw a series of boxes in a long strip.
• Enter each outcome in a box.

4•6 PROBABILITY

Check It Out

Consider the following strip graph.

red	white	red	yellow	red	red

10. Describe the first six outcomes.
11. What do you think the experiment might be?
12. Make a strip graph to show the outcomes if you roll a number cube 20 times. Compare your graphs to others' graphs.

Outcome Grids

You have seen how to use a tree diagram to show possible outcomes. Another way to show the outcomes in an experiment is to use an **outcome grid.**

MAKING OUTCOME GRIDS

Make an outcome grid to show the results of tossing two coins.

• List the outcomes of tossing a coin down the side and across the top.

2nd toss

	Head	Tail
Head		
Tail		

1st toss

• Fill in the outcomes.

2nd toss

	Head	Tail
Head	H,H	H,T
Tail	T,H	T,T

1st toss

Once you have completed the outcome grid, it is easy to count target outcomes and determine probabilities.

Check It Out

13. Make an outcome grid to show the two-digit outcomes when spinning the spinner twice. The first spin produces the first digit. The second spin produces the second digit.

14. What is the probability of getting a number divisible by 11 when you spin the spinner in item 13 twice?

Probability Line

You know that the probability of an event is a number from 0 to 1. One way to show probabilities and how they relate to each other is to use a **probability line.** The following probability line shows the possible ranges of probability values.

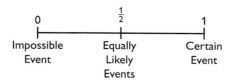

The line shows that events that are certain to happen have a probability of 1. Such an event is the probability of getting a number between 0 and 7 when a standard number cube is rolled. An event that cannot happen has a probability of zero. The probability of getting an 8 when spinning a spinner that shows 0, 2, and 4 is 0. Events that are equally likely, such as getting a head or a tail when you toss a coin, have a probability of $\frac{1}{2}$.

4·6 PROBABILITY

SHOWING PROBABILITY ON A PROBABILITY LINE

If you toss two coins, you can use a number line to show the probabilities of getting two tails and of getting a tail and a head.

• Draw a number line and label it from 0 to 1.

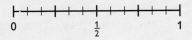

• Calculate the probabilities of the given events and show them on the probability line.

> From the outcome grid on page 228, you can see that you get two tails once and a head and a tail twice. $P(2 \text{ tails}) = \frac{1}{4}$ and $P(\text{head and tail}) = \frac{2}{4} = \frac{1}{2}$. The probabilities are shown on the following probability line.

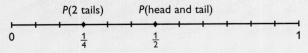

Check It Out

Draw a probability line. Then plot the following:

15. the probability of drawing a red marble from a bag of red marbles
16. the probability of rolling a 5 or a 6 on one roll of a die
17. the probability of getting both heads or both tails when flipping a coin twice
18. the probability of drawing a white marble from a bag of red marbles

Birthday Surprise

How likely do you think it is that two people in your class have the same birthday? With 365 days in a year, you might think the chances are very slim. After all, the probability that a person is born on any given day is $\frac{1}{365}$, or about 0.3%.

Try taking a survey. Ask your classmates to write their birthdays on separate slips of paper. Don't forget to write your birthday, too. Collect the slips and see if any birthdays match. If you get a match, note the number of slips you have to go through before a match occurs. If you don't get a match, try again with a different group of people.

It might surprise you to learn that in a group of 23 people, the chances that two share the same birthday is just a slight bit more than 50%. With 30 people, the likelihood increases to 71%. And with 50 people, you can be 97% sure that two of them were born on the same day.

Dependent and Independent Events

If you roll a number cube and toss a coin, the result of one does not affect the other. We call these events **independent events.** To find the probability of getting a 4 and then a tail, you can find the probability of each event and then multiply. The probability of getting a 4 on a roll of the number cube is $\frac{1}{6}$ and the probability of getting a tail is $\frac{1}{2}$. So the probability of getting a 4 and a tail is $\frac{1}{6} \times \frac{1}{2} = \frac{1}{12}$.

Suppose you have seven yellow and three white tennis balls in a bag. The probability that you get a white tennis ball if you choose a ball at random is $\frac{3}{10}$. Once you have taken a white tennis ball out, however, there are only nine balls left, two of which are white. So the probability that a friend gets a white tennis ball once you have drawn one out is $\frac{2}{9}$. These events are called **dependent events,** because the probability of one depends on the other.

In the case of dependent events, you still multiply to get the probability of both events happening. So the probability that your friend gets a white tennis ball and you also get one is $\frac{3}{10} \times \frac{2}{9} = \frac{1}{15}$.

To find the probability of independent or dependent events:
- Find the probability of the first event.
- Find the probability of the second event.
- Find the product of the two probabilities.

Check It Out

19. Find the probability of getting a 5 and an even number if you roll two number cubes. Are the events dependent or independent?
20. You draw a cookie from a bag containing eight chocolate cookies and 12 oatmeal ones. What is the probability that you get two oatmeal cookies? Are the events dependent or independent?

Sampling With and Without Replacement

If you draw a card from a deck of 20 cards each containing one of the numbers 1–20, the probability that the card you draw is an even number is $\frac{10}{20}$, or $\frac{1}{2}$. If you put the card back in the deck and draw another card, the probability that the card you draw is an even number is still $\frac{1}{2}$, and the events are independent. This is called **sampling with replacement.**

If you do not put the card back in, the probability of drawing an even number the second time will depend on what you drew the first time. If you drew an even number, there will be only nine even numbers left out of 19 cards, so the probability of drawing a second even number will be $\frac{9}{19}$. In sampling without replacement, the events are dependent.

Check It Out

21. You draw a letter from cards containing the letters of the name *Mississippi*. Then you replace that card, shuffle them, and draw again. What is the probability that you draw the letter *i* twice in a row?
22. If you do not replace the card from item 21, what probability occurs?

Flavor of the Week

Marketing research often involves the use of surveys and product sampling to help predict how well certain products will do. The results of this process might then be used to determine what part of the population is likely to buy these new flavors.

A company wishing to introduce three new flavors of ice cream—Banana Bonanza, Raspberry Rush, and Kiwi Kiss—handed out samples of each flavor to 300 seventh graders. Among other things, they were then asked which flavor or flavors they liked and which they didn't like. The results were as follows: 145 liked Raspberry Rush, 25 liked Kiwi Kiss, 15 liked both Raspberry Rush and Kiwi Kiss, 10 liked both Banana Bonanza and Kiwi Kiss, 100 liked both Banana Bonanza and Raspberry Rush, and 5 liked all three flavors.

Use the information provided above to predict how many out of ten seventh graders at your school should prefer Banana Bonanza. See Hot Solutions for answer. Then actually survey ten seventh graders about the flavor they think they would prefer.

4·6 EXERCISES

You roll a six-sided number cube numbered 1 through 6. Find each probability as a fraction, decimal, ratio, and percent in items 1–2.

1. $P(4 \text{ or } 5)$

2. $P(\text{even number})$

3. If you draw a card from a regular deck of 52 cards, what is the probability of getting an ace? Is this experimental or theoretical probability?

4. If you toss a thumbtack 48 times and it lands up 15 times, what is the probability of it landing up again on the next toss? Is this experimental or theoretical probability?

5. Draw a probability line to show the probability of getting a number less than 1 when you are rolling a six-sided number cube numbered 1 through 6.

6. Make a strip graph to show the following outcomes from rolling a six-sided number cube: 2, 3, 6, 4, 5, 3, 4, 1, 2.

7. Make an outcome grid to show the outcomes of spinning two spinners containing the numbers 1–5.

8. Find the probability of drawing a red ball and a white ball from a bag of 16 red and 24 white balls if you replace the ball between drawings.

9. Find the probability of drawing a red ball and a white ball (item 8) if you do not replace the ball between drawings.

10. In which of items 8 and 9 are the events independent?

What have you learned?

You can use the problems and the list of words that follow to see what you have learned in this chapter. You can find out more about a particular problem or word by referring to the boldfaced topic number (for example, **6•2**).

Problem Set

1. Mr. Chan took a survey of his class. He gave each student a number and then put duplicate numbers in a paper bag. He drew ten numbers from the bag without peeking. He then surveyed those ten students. Was this a random sample? **4•1**

2. A survey asked, "Are you a responsible citizen who has registered to vote?" Rewrite the question so it is not biased. **4•1**

3. What kind of graph can be used to show how often something occurs? **4•2**

4. On a circle graph, how many degrees must be in a sector to show 25%? **4•2**

To answer items 5–7, use the following bar graph, which shows the inventions people said they could not live without. **4•2**

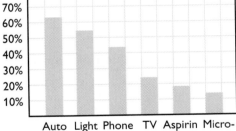

INVENTIONS PEOPLE SAY THEY CAN'T LIVE WITHOUT

5. Which inventions did more than 50% of the people surveyed say they could not do without?

6. What percent of the people did not feel they could do without the microwave?

7. What kind of graph shows the data divided into quartiles? **4•2**

8. Is the following statement true or false? Explain. A histogram uses X's to show how many times something occurs. **4•2**

Use the two graphs
to answer items
9 and 10. **4•3**

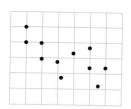
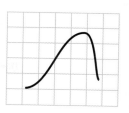

9. Describe the correlation on the left.
10. Describe the distribution on the right.

11. Find the mean, median, mode, and range of the numbers 18, 23, 18, 15, 20, and 17. **4•4**

12. Of the mean, median, and mode, which must be a member of the set of data? **4•4**

13. $C(9, 2) = ?$ **4•5**

14. $\frac{10!}{8!} = ?$ **4•5**

15. What is the probability of drawing a 4 from a deck of cards? **4•6**

---

hot **words**

WRITE DEFINITIONS FOR THE
FOLLOWING WORDS.

average   **4•4**
bimodal distribution   **4•3**
box plot   **4•2**
circle graph   **4•2**
combination   **4•5**
correlation   **4•3**
dependent events   **4•6**
distribution   **4•3**
double-bar graph   **4•2**
event   **4•6**
experimental probability   **4•6**
factorial   **4•5**
flat distribution   **4•3**
histogram   **4•2**
independent events   **4•6**
leaf   **4•2**
line graph   **4•2**
mean   **4•4**
median   **4•4**
mode   **4•4**

normal distribution   **4•3**
outcome   **4•6**
outcome grid   **4•6**
percent   **4•2**
permutation   **4•5**
population   **4•1**
probability   **4•6**
probability line   **4•6**
random sample   **4•1**
range   **4•4**
sample   **4•1**
sampling with replacement   **4•6**
scatter plot   **4•3**
skewed distribution   **4•3**
spinner   **4•5**
stem   **4•2**
stem-and-leaf plot   **4•2**
strip graph   **4•6**
survey   **4•1**
table   **4•1**
tally marks   **4•1**
theoretical probability   **4•6**
tree diagram   **4•5**

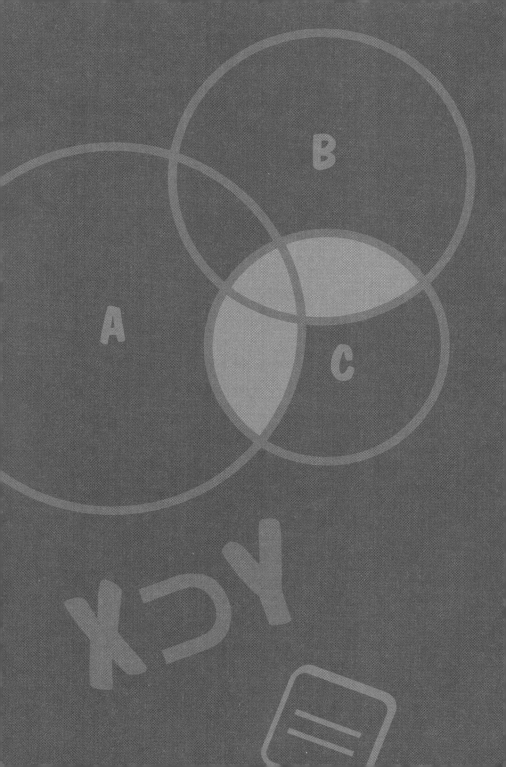

Logic

What do you already know?

You can use the problems and list of words below to see what you already know about this chapter. The answers to the problems are in Hot Solutions at the back of the book, and the definitions of the words are in Hot Words at the front of the book. You can find out more about a particular problem or word by referring to the boldfaced topic number (for example, **5•2**).

Problem Set

Tell whether each statement is true or false.

1. In an if/then statement, the *if* part is the hypothesis. **5•1**
2. If a conditional statement is true, then its related converse is always true. **5•1**
3. If a conditional statement is true, then its related contrapositive is false. **5•1**
4. The negation of "It's summer" is "It's winter." **5•1**
5. To form the union of two sets, you combine all the elements in both sets. **5•3**
6. To form the intersection of two sets, you make a set of the elements that are common to both sets. **5•3**
7. A counterexample shows that a statement is true. **5•2**
8. A counterexample agrees with the hypothesis of a conditional but not with the conclusion. **5•2**
9. Every set is a subset of itself. **5•3**

Write each conditional in if/then form. **5•1**

10. The streetlights are on after 7 P.M.
11. An acute angle has a measure greater than 0° but less than 90°.

Write the converse of each conditional statement. **5•1**

12. If $a = 4$, then $a^2 = 16$.
13. If the bus is running on schedule, then I arrive at school on time.

Write the negation of each statement. **5•1**

14. The street is closed for repairs.
15. These two lines are not perpendicular.

Write the inverse of each conditional statement. **5•1**
16. If $x + 7 = 12$, then $x = 5$.
17. If you do all your homework, then you will receive a good grade.

Write the contrapositive of each conditional statement. **5•1**
18. If you watch too much television, then you become lazy.

19. If you flipped this coin, then the outcome was a head or a tail.

Find a counterexample that shows that each of these statements is false. **5•2**
20. July 4 is the only national holiday in the summer.
21. The number 3 has only odd multiples.

22. Find all the subsets of $\{a, b, c\}$. **5•3**

Find the union of each pair of sets. **5•3**
23. $\{11, 12, 13\} \cup \{14, 15\}$
24. $\{11, 12, 13\} \cup \{12, 14\}$

Find the intersection of each pair of sets. **5•3**
25. $\{m, n, o, p\} \cap \{q, r\}$
26. $\{5, 10, 15, 20\} \cap \{5, 15, 25, 35\}$

Use the Venn diagram to answer items 27–30. **5•3**
27. List the elements in set A.
28. List the elements in set B.
29. Find $A \cup B$.
30. Find $A \cap B$.

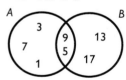

CHAPTER 5

ℎ𝑜𝑡**words**

converse **5•1**
counterexample **5•2**
contrapositive **5•1**
intersection **5•3**

inverse **5•1**
union **5•3**
Venn diagram **5•3**

5·1 If/Then Statements

Conditional Statements

A *conditional* is a statement that you can express in if/then form. The *if* part of a conditional is the *hypothesis,* and the *then* part is the *conclusion.* Often you can rewrite a statement that contains two related ideas as a conditional in if/then form. Do this by making one of the ideas the hypothesis and the other the conclusion.

Statement: Students in the honor society have high grades. The conditional in if/then form:

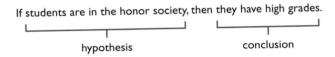

If students are in the honor society, then they have high grades.

hypothesis · conclusion

FORMING CONDITIONAL STATEMENTS

Write this conditional in if/then form:

People who arrive on time get good seats.

- Find the two ideas.
  - (1) people arrive on time
  - (2) people get good seats
- Decide which idea will be the hypothesis and which will be the conclusion.

  Hypothesis: people arrive on time

  Conclusion: people get good seats
- Place the hypothesis in the *if* part and the conclusion in the *then* part. If necessary, add words so that your sentence makes sense.

If people arrive on time, then they get good seats.

Check It Out

Write each conditional statement in if/then form.
1. Two segments with the same length are congruent.
2. An integer that ends with 0 is a multiple of 10.

Converse of a Conditional

When you switch the hypothesis and conclusion in a
conditional statement, you form a new statement called the
converse.

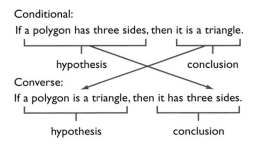

Conditional:
If a polygon has three sides, then it is a triangle.

hypothesis   conclusion

Converse:
If a polygon is a triangle, then it has three sides.

hypothesis   conclusion

The converse of a conditional may or may not have the same
truth value as the conditional on which it is based.

 Check It Out

Write the converse of each conditional.
3. If you are in the seventh grade, then you study
   algebra.
4. If you add 3 and 4, then you get a sum of 7.

Negations and the Inverse of a Conditional

A *negation* of a given statement has the opposite truth value of the given statement. That means that if the given statement is true, the negation is false; if the given statement is false, the negation is true.

> Statement: Five is an odd number. (True)
> Negation: Five is not an odd number. (False)

> Statement: A square has three sides. (False)
> Negation: A square does not have three sides. (True)

When you negate the hypothesis and conclusion of a conditional, you form a new statement called the **inverse**.

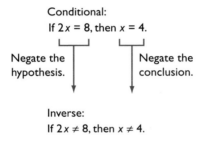

Conditional:
If $2x = 8$, then $x = 4$.

Negate the hypothesis.   Negate the conclusion.

Inverse:
If $2x \neq 8$, then $x \neq 4$.

The inverse of a conditional may or may not have the same truth value as the conditional.

Check It Out

Write the negation of each statement.
5. 5 is greater than 4.
6. We will go to the beach this summer.

Write the inverse of each conditional.
7. If an integer ends with 0, then it is a multiple of 10.
8. If you live in San Diego, then you live in California.

Contrapositive of a Conditional

You form the **contrapositive** of a conditional when you negate the hypothesis and conclusion, and then switch them.

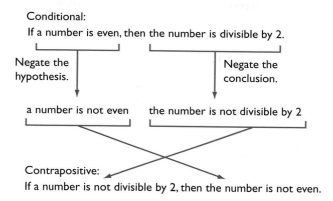

Conditional:
If a number is even, then the number is divisible by 2.

Negate the hypothesis.

Negate the conclusion.

a number is not even the number is not divisible by 2

Contrapositive:
If a number is not divisible by 2, then the number is not even.

The contrapositive of a conditional statement has the same truth value as the conditional.

Check It Out

Write the contrapositive of each conditional.

9. If a number is divisible by 6, then it is divisible by 3.

10. If two lines are parallel, then they do not intersect.

Rapunzel, Rapunzel, Let Down Your Hair

Do you think there are two people in the United States with exactly the same number of hairs on their head? In fact, you can prove that there are. Not by counting the hairs on everyone's head, but by logic.

Consider these statements:

A. At the maximum, there are approximately 150,000 hairs on the human scalp.

B. The population of the United States is greater than 150,000.

Because statements a and b are both true, there are two people in the United States with exactly the same number of hairs on their head.

Here is how to think about this. If you did count hairs, the first 150,001 people could each have a different number. Person 1 could have 1 hair; person 2, 2 hairs, and so on to 150,000. But person 150,001 would have to have a number of hairs between 1 and 150,000 and this would be a duplicate of one of the heads you already counted.

Can you prove that there are two people in your town with the same number of hairs on their head? See Hot Solutions for answer.

5·1 EXERCISES

Write each conditional in if/then form.
1. A number whose last digit is 4 is an even number.
2. A right triangle has one right angle.
3. A square has four equal sides.
4. The lifeguard had to pass a swimming test.
5. Cleve works every Saturday.
6. Every student in that class is getting promoted.

Write the converse of each conditional.
7. If $x = 3$, then $x + 4 = 7$.
8. If a triangle has three equal sides, then it is equilateral.
9. If two lines intersected, then four angles were formed.

Write the negation of each statement.
10. Twenty is a multiple of 4.
11. A pentagon has five sides.
12. All the buildings have elevators.

Write the inverse of each conditional.
13. If $x + 8 = 13$, then $x = 5$.
14. If a rectangle has equal sides, then it is a square.
15. If my ticket number is chosen, then I will win a prize.

Write the contrapositive of each conditional.
16. If $\frac{x}{6} = 5$, then $x = 30$.
17. If a triangle has at least two equal sides, then it is isosceles.
18. If you subtracted 3 from 9, then you got 6 for an answer.

For each conditional, write the converse, inverse, and contrapositive.
19. If the sides of a triangle are 3 ft, 4 ft, and 5 ft, then the perimeter of the triangle is 12 ft.
20. If you did all your homework, then you passed the course.

5·2 Counterexamples

Counterexamples

Any if/then statement is either true or false. One way to decide whether a statement is false is to find just one example that agrees with the hypothesis but not with the conclusion. Such an example is a **counterexample.** When reading the conditional below, you may be tempted to think that it is true.

If a number is prime, then it is an odd number.

The statement is false, however, because there is a counterexample—the number 2. The number 2 agrees with the hypothesis (it is prime), but it does not agree with the conclusion (2 is an even number).

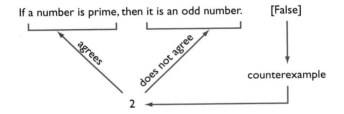

Check It Out

Tell if each statement and its converse are true or false. If a statement is false, give a counterexample.

1. Statement: If an angle has a measure of 15°, then it is acute.

   Converse: If an angle is acute, then it has a measure of 15°.

2. Statement: If two numbers are odd, then their product is odd.

   Converse: If the product of two numbers is odd, then the numbers are odd.

5·2 EXERCISES

Find a counterexample that shows that each of these statements is false.

1. If an angle is obtuse, then it has a measure greater than 100°.
2. If a figure has four sides, then the figure is a trapezoid.
3. If $a \times b$ is an even number, then a and b are both even.
4. If two rectangles have the same perimeter, then they have the same area.

Tell whether each conditional is true or false. If false, give a counterexample.

5. If the temperature is 212°F or above, then water boils.
6. If you draw a line through a rectangle, then you form two smaller rectangles.
7. If $x^2 = 25$, then $x = 5$.
8. If a whole number is between 24 and 28, then it is composite.

Tell if each conditional (A) and its converse (B) are true or false. If false, give a counterexample.

9. A. If $9x = 36$, then $x = 4$.
   B. If $x = 4$, then $9x = 36$.
10. A. If two segments are 8 cm long, then they are congruent.
    B. If two segments are congruent, then they are 8 cm.

Tell if each conditional (A) and its inverse (B) are true or false. If false, give a counterexample.

11. A. If $n = 4$, then $n + 9 = 13$.
    B. If $n \neq 4$, then $n + 9 \neq 13$.
12. A. If you add 7 and 9, then you get a sum of 16.
    B. If you do not add 7 and 9, then you do not get a sum of 16.

Tell if each conditional (A) and its contrapositive (B) are both true or false. If false, give a counterexample.

13. A. If a rectangle has a length of 5 ft and a width of 3 ft, then its area is 15 ft^2.
    B. If a rectangle does not have an area of 15 ft^2, then the rectangle does not have a length of 5 ft and a width of 3 ft.
14. A. If a number is odd, then it is divisible by 3.
    B. If a number is not divisible by 3, then it is not odd.

15. Write your own conditional, then its converse, inverse, and contrapositive.

5·2 EXERCISES

5·3 Sets

Sets and Subsets

A *set* is a collection of objects. Each object is called a *member* or *element* of the set. Sets are often named with capital letters.

$A = \{1, 2, 3\}$ $B = \{x, y, z\}$

When a set has no elements, it is an *empty* set. You write { } or $\emptyset$ to indicate the empty set.

When all the elements of a set are also elements of another set, the first set is a *subset* of the other set.

$\{1, 3\}$ is a subset of $\{1, 2, 3\}$.

$\{1, 3\} \subset \{1, 2, 3\}$ ($\subset$ is the subset symbol.)

Remember that every set is a subset of itself and that the empty set is a subset of every set.

Check It Out

Tell whether each statement is true or false.

1. $\{4\} \subset \{$odd numbers$\}$ 2. $\emptyset \subset \{2, 4\}$

Find all the subsets of each set.

3. $\{4, 8\}$ 4. $\{m\}$

Union of Sets

To find the **union** of two sets, you create a new set with all the elements from the two sets.

$R = \{1, 3, 5\}$ $T = \{2, 4, 6\}$

$R \cup T = \{1, 2, 3, 4, 5, 6\}$ ($\cup$ is the union symbol.)

When the sets have elements in common, list the common elements only once in the intersection.

$P = \{7, 8, 9\}$ $Q = \{6, 9, 12\}$

$P \cup Q = \{6, 7, 8, 9, 12\}$

Check It Out
Find the union of each pair of sets.
5. $\{3,6\} \cup \{8,10\}$ 6. $\varnothing \cup \{x, y\}$

Intersection of Sets

You find the **intersection** of two sets by creating a new set that contains all the elements that are common to both sets.

$$J = \{\textcircled{3}, \textcircled{6}, 9, 12\}$$
$$K = \{\textcircled{3}, 4, 5, \textcircled{6}\}$$
$$J \cap K = \{3, 6\}$$

If the sets have no elements in common, the intersection is the empty set ($\varnothing$).

 Check It Out
Find the intersection of each pair of sets.
7. $\{7, 14\} \cap \{14\}$ 8. $\{b, c\} \cap \{d, e\}$

Venn Diagrams

A **Venn diagram** shows you how the elements of two or more sets are related.

$A = \{1, 2, 3\}$
$B = \{a, b, c\}$
$A \cup B = \{1, 2, 3, a, b, c\}$

The separate circles for A and B tell you that the sets have no elements in common. That means that $A \cap B = \varnothing$.

When the circles in a Venn diagram overlap, the overlapping part contains the elements that are common to both sets. This diagram shows some sets of attribute shapes.

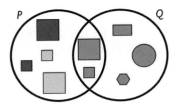

$P = \{\text{squares}\}$
$Q = \{\text{red shapes}\}$

The overlapping parts of P and Q contain shapes that have the attributes of both sets, or $P \cap Q = \{\text{red squares}\}$.

When there are more complex Venn diagrams, you have to look carefully to identify the overlapping parts and see which elements of the sets are in those parts. In this diagram, H overlaps J, H overlaps K, and J overlaps K. The shaded part of the diagram shows where all three sets overlap one another.

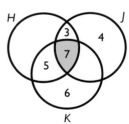

$H = \{3, 5, 7\}$ $H \cup J = \{3, 4, 5, 7\}$
$J = \{3, 4, 7\}$ $H \cup K = \{3, 5, 6, 7\}$
$K = \{5, 6, 7\}$ $J \cup K = \{3, 4, 5, 6, 7\}$
$H \cap J = \{3, 7\}$
$H \cap K = \{5, 7\}$
$J \cap K = \{7\}$
$H \cup J \cup K = \{3, 4, 5, 6, 7\}$

Where all three sets overlap, you can see that $H \cap J \cap K = \{7\}$.

 Check It Out
List the elements in:
  9. set X.
 10. $X \cap Z$.
 11. $Y \cup Z$.
 12. $Y \cap X$.
 13. $X \cap Y \cap Z$.

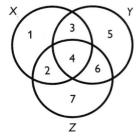

5·3 EXERCISES

Tell whether each statement is true or false.
1. $\{2, 4, 5\} \subset \{$even numbers$\}$
2. $\{1, 2, 3\} \subset \{1, 2, 3\}$
3. $\varnothing \subset \{2, 4, 5\}$
4. $\{3, 5, 7\} \subset \{$odd numbers$\}$
5. List all the subsets of $\{2, 4, 6\}$.

Find the union of each pair of sets.
6. $\{1, 2\} \cup \{5, 6\}$
7. $\{m, n\} \cup \{$n, p$\}$
8. $\{c, a, n, d, l, e\} \cup \{h, a, n, d, l, e\}$
9. $\{2, 4, 6, 8\} \cup \{4, 6\}$

Find the intersection of each pair of sets.
10. $\{2, 4, 6\} \cap \{5, 6, 7, 8\}$
11. $\{7, 14, 21\} \cap \{6, 12, 18\}$
12. $\{c, a, n, d, l, e\} \cap \{h, a, n, d, l, e\}$
13. $\varnothing \cap \{3\}$

Use the Venn diagram below to answer items 14–17.

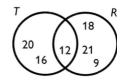

14. List the elements of set T.
15. List the elements of set R.
16. Find $T \cup R$.
17. Find $T \cap R$.

Use the Venn diagram below to answer items 18–25.

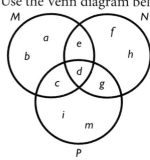

18. List the elements of set M.
19. List the elements of set N.
20. Find P.
21. Find $M \cup N$.
22. Find $N \cup P$.
23. Find $M \cap P$.
24. Find $M \cup P$.
25. Find $M \cap N \cap P$.

What have you learned?

You can use the problems and list of words below to see what you have learned in this chapter. You can find out more about a particular problem or word by referring to the boldfaced topic number (for example, **5•2**).

Problem Set

Tell whether each statement is true or false.

1. In an if/then statement, the *then* part is the conclusion. **5•1**
2. To form the converse of a conditional, you negate the hypothesis and the conclusion. **5•1**
3. When you form the contrapositive of a conditional, you negate the hypothesis and interchange it with the conclusion. **5•1**
4. The negation of "It's summer" is "It's not summer." **5•1**
5. To form the intersection of two sets, you make a set of the elements that the two sets have in common. **5•3**
6. When you form the union of two sets, you combine the elements in the two sets. **5•3**
7. A counterexample shows that a statement is false. **5•2**
8. The empty set is a subset of every set. **5•3**

Write each conditional in if/then form. **5•1**

9. It always rains on Thursday.
10. An obtuse angle has a measure greater than 90° but less than 180°.

Write the converse of each conditional statement. **5•1**

11. If $n = 5$, then $3 \times n = 15$.
12. If the snow accumulation is 12 in., then school is canceled.

Write the negation of each statement. **5•1**

13. Baseball season begins in April.
14. That triangle does not appear to be scalene.

Write the inverse of each conditional statement. **5•1**

15. If $c + 6 = 10$, then $c = 4$.
16. If you score a 95 on the next test, then you will get an A.

Write the contrapositive of each conditional statement. **5•1**
17. If you are over 48 in. tall, then you pay full fare.

18. If a figure is a rectangle, then its area formula is $A = lw$.

Find a counterexample that shows that each of these statements is false. **5•2**
19. The number 12 has only even factors.
20. Saturday is the only day of the week that begins with the letter *s*.
21. Find all the subsets of $\{3, 6, 9\}$. **5•3**

Find the union of each pair of sets. **5•3**
22. $\{10, 11, 12\} \cup \{13, 14\}$
23. $\{10, 11, 12\} \cup \{12, 13\}$

Find the intersection of each pair of sets. **5•3**
24. $\{m, o, p, e, d\} \cap \{m, o, d, e\}$
25. $\{3, 6, 9, 12\} \cap \{4, 8, 16, 32\}$

Use the Venn diagram to answer items 26–30. **5•3**
26. List the elements in set *A*.
27. List the elements in set *C*.
28. Find $A \cup B$.
29. Find $B \cap C$.
30. Find $A \cap B \cap C$.

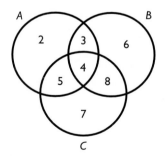

WHAT HAVE YOU LEARNED?

WRITE DEFINITIONS FOR THE FOLLOWING WORDS.

hot **words**

converse **5•1**
counterexample **5•2**
intersection **5•3**

inverse **5•1**
union **5•3**
Venn diagram **5•3**

contrapositive **5•1**

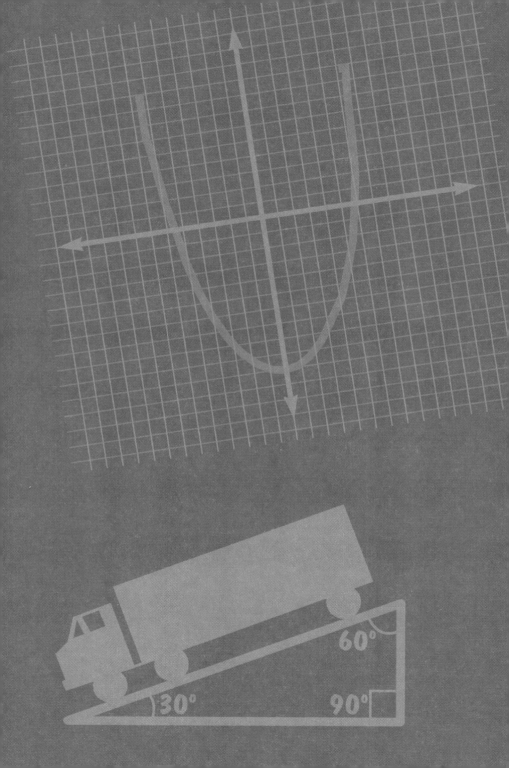

Algebra

What do you already know?

You can use the problems and the list of words that follow to see what you already know about this chapter. The answers to the problems are in Hot Solutions at the back of the book, and the definitions of the words are in Hot Words at the front of the book. To find out more about a particular problem or word, refer to the boldfaced topic number (for example, **6•2**).

Problem Set

Write an equation for each sentence. **6•1**

1. If 5 is subtracted from twice a number, the result is 3 more than the number.
2. 6 times the sum of a number and 2 is 4 less than twice the number.

Factor out the greatest common factor in each expression. **6•2**

3. $4x + 24$
4. $9n - 12$

Simplify each expression. **6•2**

5. $7a + 6b - a - 4b$
6. $3(2n - 1) - (n + 4)$
7. Find the distance traveled by a jogger who jogs at 4 mi/hr for $2\frac{1}{2}$ hr. Use the formula $d = rt$. **6•3**

Solve each equation. Check your solution. **6•4**

8. $x + 3 = 9$
9. $\frac{y}{3} = -6$
10. $4x - 3 = 17$
11. $\frac{y}{5} - 3 = 4$
12. $7n - 8 = 4n + 7$
13. $y - 8 = 4y + 4$
14. $3(2n - 3) = 4n + 7$
15. $8x - 2(3x - 2) = 6(x + 2)$

Use a proportion to solve each problem. **6•5**

16. In a class, the ratio of boys to girls is $\frac{2}{3}$. If there are 10 boys in the class, how many girls are there?
17. A map is drawn using a scale of 200 mi to 1 cm. The distance between two cities is 1400 mi. How far apart are the two cities on the map?

Solve each inequality. Graph the solution. **6•6**

18. $x + 4 < 3$ 19. $3x + 2 \geq 11$

Locate each point on the coordinate plane and tell in which quadrant or on which axis it lies. **6•7**

20. $A(-3, 4)$ 21. $B(3, 0)$ 22. $C(0, -2)$ 23. $D(2, 3)$

24. Find the slope of the line that contains the points $(4, -2)$ and $(-2, 8)$.

Determine the slope and the y-intercept from the equation of each line. Graph the line. **6•8**

25. $y = \frac{-2}{3}x + 3$ 26. $y = -2$

27. $x + 3y = -3$ 28. $6x - 3y = 0$

Write the equation of the line that contains the given points. **6•8**

29. $(-2, -5)$ and $(5, 2)$

30. $(4, 7)$ and $(-2, -2)$

CHAPTER 6

hot **words**

additive inverse **6•4**
associative property **6•2**
axes **6•7**
commutative property **6•2**
cross product **6•5**
difference **6•1**
distributive property **6•2**
equation **6•1**

equivalent **6•1**
equivalent expression **6•2**
expression **6•1**
formula **6•3**
horizontal **6•7**
inequality **6•6**
like terms **6•2**
order of operations **6•3**
ordered pair **6•7**
origin **6•7**
perimeter **6•3**
point **6•7**
product **6•1**

proportion **6•5**
quadrant **6•7**
quotient **6•1**
rate **6•5**
ratio **6•5**
slope **6•8**
solution **6•4**
sum **6•1**
term **6•1**
variable **6•1**
vertical **6•7**
x-axis **6•7**
y-axis **6•7**
y-intercept **6•8**

WHAT DO YOU KNOW?

6·1 Writing Expressions and Equations

Expressions

In mathematics, often the value of a certain number may be unknown. A **variable** is a symbol, usually a letter, that is used to represent an unknown number. Some commonly used variables are:

$$x \quad n \quad y \quad a \quad ?$$

A **term** can be a number, a variable, or a number and variable combined by multiplication or division. Some examples of terms are:

$$w \quad 5 \quad 3x \quad \frac{y}{8}$$

An **expression** can be a term or a collection of terms separated by addition or subtraction signs. Some expressions, with the number of terms, are listed in the table below.

Expression	Number of Terms	Description
$5y$	1	a number multiplied by a variable
$6z + 4$	2	terms separated by a +
$3x + 7a - 5$	3	
$\frac{9xz}{y}$	1	all multiplication and division; no + symbol

 Check It Out

Count the number of terms in each expression.
1. $3n + 8$
2. $4xyz$
3. $6ab - 2c - 5$
4. $2(x - 5) + 12$

Writing Expressions Involving Addition

To write an expression, you often have to interpret a written phrase. For example, the phrase "4 added to some number" can be written as the expression $x + 4$, where the variable x represents the unknown number.

Notice that the words "added to" indicate that the operation between 4 and the number is to be addition. Other words and phrases that indicate addition are "more than," "plus," and "increased by." One other word that indicates addition is **sum**. The sum of two terms is the result of adding them together.

Here are some common phrases and their corresponding expressions.

Phrase	Expression
3 more than some number	$n + 3$
a number increased by 7	$x + 7$
9 plus some number	$9 + y$
the sum of a number and 6	$n + 6$

Check It Out

Write an expression for each phrase.
5. a number added to 3
6. the sum of a number and 9
7. some number increased by 5
8. 4 more than some number

Writing Expressions Involving Subtraction

The phrase "4 subtracted from some number" can be written as the expression $x - 4$, where the variable x represents the unknown number. Notice that the words "subtracted from" indicate that the operation between the number and 4 is to be subtraction.

Some other words and phrases that indicate subtraction are "less than," "minus," and "decreased by." One other word that indicates subtraction is **difference**. The difference between two terms is the result of subtracting them.

In a subtraction expression, the order of the terms is very important. You have to know which term is being subtracted and which is being subtracted from. To help interpret the phrase "6 less than a number," replace "a number" with 10. What is 6 less than 10? The answer is 4, which is $10 - 6$, not $6 - 10$. The phrase translates to the expression $x - 6$, not $6 - x$.

Some common phrases and their corresponding expressions are listed.

Phrase	Expression
5 less than some number	$n - 5$
a number decreased by 8	$x - 8$
7 minus some number	$7 - y$
the difference between a number and 2	$n - 2$

Check It Out

Write an expression for each phrase.
9. a number subtracted from 10
10. the difference between a number and 7
11. some number decreased by 5
12. 8 less than some number

Writing Expressions Involving Multiplication

The phrase "4 multiplied by some number" can be written as the expression $4x$, where the variable x represents the unknown number. Notice that the words "multiplied by" indicate that the operation between the number and 4 is to be multiplication.

Some other words and phrases that indicate multiplication are "times," "twice," and "of." "Twice" is used to mean "2 times." "Of" is used primarily with fractions and percents. One other word that indicates multiplication is **product.** The product of two terms is the result of multiplying them.

Some common phrases and their corresponding expressions are listed.

Phrase	Expression
5 times some number	$5a$
twice a number	$2x$
one-fourth of some number	$\frac{1}{4}y$
the product of a number and 8	$8n$

Check It Out

Write an expression for each phrase.
13. a number multiplied by 6
14. the product of a number and 4
15. 75% of some number
16. 10 times some number

Writing Expressions Involving Division

The phrase "4 divided by some number" can be written as the expression $\frac{4}{x}$, where the variable x represents the unknown number. Notice that the words "divided by" indicate that the operation between the number and 4 is to be division.

Some other words and phrases that indicate division are "ratio of," and "divide." One other word that indicates division is **quotient.** The quotient of two terms is the result of one being divided by the other.

Some common phrases and their corresponding expressions are listed.

Phrase	Expression
the quotient of 20 and some number	$\frac{20}{n}$
a number divided by 6	$\frac{x}{6}$
the ratio of 10 and some number	$\frac{10}{y}$
the quotient of a number and 5	$\frac{n}{5}$

Check It Out

Write an expression for each phrase.

17. a number divided by 3
18. the quotient of 12 and a number
19. the ratio of 30 and some number
20. the quotient of some number and 7

Writing Expressions Involving Two Operations

To translate the phrase "2 added to the product of 3 and some number" to an expression, first realize that "2 added to" means "something" + 2. That "something" is "the product of 3 and some number," which is $3x$, since "product" indicates multiplication. Thus the expression can be written as $3x + 2$.

Phrase	Expression	Think
2 less than the quotient of a number and 5	$\frac{x}{5} - 2$	"2 less than" means "something" − 2; "quotient" indicates division.
5 times the sum of a number and 3	$5(x + 3)$	Write the sum inside parentheses so that the entire sum is multiplied by 5.
3 more than 7 times a number	$7x + 3$	"3 more than" means "something" + 3; "times" indicates multiplication.

Check It Out

Translate each phrase to an expression.
21. 8 less than the product of 5 and a number
22. 4 subtracted from the quotient of 2 and a number
23. twice the difference between a number and 10

Orphaned Whale Rescued

On January 11, 1997, an orphaned baby gray whale arrived at an aquarium in California. Rescue workers named her J. J. She was three days old, weighed 1,600 lb, and was desperately ill.

Soon her caretakers had her sucking from a tube attached to a thermos. By February 7, on a diet of whale milk formula, she weighed 2,378 pounds. J. J. was gaining 20 to 30 lb a day!

An adult gray whale weighs approximately 35 tons, but the people at the aquarium knew they could release J. J. once she had a solid layer of blubber—when she weighed about 9,000 lb.

Write an equation that shows J. J. is 2,378 lb now, and needs to gain 25 lb per day for some number of days until she weighs 9,000 lb. See Hot Solutions for answer.

6·1 WRITING EXPRESSIONS

Equations

An expression is a phrase; an **equation** is a sentence. An equation indicates that two expressions are **equivalent,** or *equal.* The symbol used in an equation is the equals sign, "=."

To translate the sentence "3 less than the product of a number and 4 is the same as 6 more than the number" to an equation, first identify the words that indicate "equals." In this sentence, "equals" is indicated by "is the same as." In other sentences "equals" may be "is," "the result is," "you get," or just "equals."

Once you have identified the =, you can then translate the phrase that comes before the = and write the expression on the left side. Then you translate the phrase that comes after the = and write the expression on the right side.

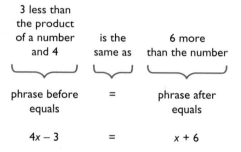

3 less than the product of a number and 4	is the same as	6 more than the number
phrase before equals	=	phrase after equals
$4x - 3$	=	$x + 6$

 Check It Out

Write an equation for each sentence.

24. 6 subtracted from 3 times a number is the same as 2 more than the product of 4 and the number.

25. If 5 is added to the quotient of a number and 4, the result is 10 less than the number.

26. 2 less than 3 times a number is twice the sum of the number and 3.

6·1 EXERCISES

Count the number of terms in each expression.
1. $3x + 5$ 2. 7
3. $5x - 2y + 4z$ 4. $8n - 13$

Write an expression for each phrase.
5. 6 more than a number
6. a number added to 9
7. the sum of a number and 5
8. 4 less than a number
9. 10 decreased by some number
10. the difference between a number and 6
11. one fourth of some number
12. twice a number
13. the product of a number and 4
14. a number divided by 6
15. the ratio of 8 and some number
16. the quotient of a number and 3

Write an expression for each phrase.
17. 7 more than the product of a number and 4
18. 1 less than twice a number
19. twice the sum of 4 and a number

Write an equation for each sentence.
20. 6 more than the quotient of a number and 4 is the same as 3 less than the number.
21. If 5 is subtracted from twice a number, the result is 11.
22. 4 times the sum of a number and 3 is 4 more than twice the number.

Select the correct response.
23. Which of the following words is used to indicate multiplication?
    A. Sum   B. Difference C. Product   D. Quotient
24. Which of the following does not indicate subtraction?
    A. Less than B. Difference C. Decreased by D. Ratio of
25. Which of the following shows "twice the sum of a number and 8"?
    A. $2(x + 8)$ B. $2x + 8$ C. $(x - 8)$ D. $2 + (x + 8)$

6·2 Simplifying Expressions

Terms

As you may remember, terms can be numbers, variables, or numbers and variables combined by multiplication or division. Some examples of terms are listed.

$$n \qquad 13 \qquad 3x \qquad x^3$$

Compare the terms 13 and $3x$. The value of $3x$ will change as the value of x changes. If $x = 2$, then $3x = 3(2) = 6$, and if $x = 3$, then $3x = 3(3) = 9$. Notice, though, that the value of 13 never changes—it remains constant. When a term contains a number only, it is called a *constant* term.

Check It Out

Decide whether each term is a constant term.

1. $5x$
2. 8
3. $4(n + 2)$
4. 3

The Commutative Property of Addition and Multiplication

The **commutative property** of addition states that the order of two terms being added may be switched without changing the result: $3 + 4 = 4 + 3$ and $x + 8 = 8 + x$. The commutative property of multiplication states that the order of two terms being multiplied may be switched without changing the result: $3(4) = 4(3)$ and $x \cdot 8 = 8x$.

The commutative property does not hold for subtraction or division. The order of the terms does affect the result: $5 - 3 = 2$, but $3 - 5 = -2$; $8 \div 4 = 2$, but $4 \div 8 = \frac{1}{2}$.

Check It Out

Rewrite each expression, using the commutative property of addition or multiplication.

5. $2x + 3$
6. $n \cdot 5$
7. $6 + 3y$
8. $5 \cdot 6$

The Associative Property of Addition and Multiplication

The **associative property** of addition states that the grouping of three terms being added does not affect the result: $(3 + 4) + 5 = 3 + (4 + 5)$ and $(x + 6) + 10 = x + (6 + 10)$. The associative property of multiplication states that the grouping of three terms being multiplied does not affect the result: $(2 \cdot 3) \cdot 4 = 2 \cdot (3 \cdot 4)$ and $5 \cdot 3x = (5 \cdot 3)x$.

The associative property does not hold for subtraction or division. The grouping of the numbers does affect the result: $(8 - 6) - 4 = -2$, but $8 - (6 - 4) = 6$; $(16 \div 8) \div 2 = 1$, but $16 \div (8 \div 2) = 4$.

Predicting Life Expectancy

How much longer are people expected to live in the future? Consider these statistics.

Year	Life Expectancy
1900	47.3
1920	54.1
1940	62.9
1960	69.7
1980	73.7
1990	75.4

This life expectancy data can be roughly described by the equation $y = 0.32x - 560$.

Why is this an unreasonable way to describe the data? See Hot Solutions for answer.

6•2 SIMPLIFYING EXPRESSIONS

Check It Out

Rewrite each expression, using the associative property of addition or multiplication.

9. $(3 + 7) + 10$
10. $(4 \cdot 2) \cdot 7$
11. $(4x + 3y) + 5$
12. $4 \cdot 5n$

The Distributive Property

The **distributive property** of addition and multiplication states that multiplying a sum by a number is the same as multiplying each addend by that number and then adding the two products. So $3(2 + 3) = (3 \cdot 2) + (3 \cdot 3)$.

How would you multiply $7 \cdot 99$ in your head? You might think, $700 - 7 = 693$. If you did, you have used the distributive property.

$7(100 - 1)$ Distribute the factor of 7 to each term inside the parentheses.

$= 7 \cdot 100 - 7 \cdot 1$ Simplify, using order of
$= 700 - 7$ operations.
$= 693$

The distributive property does not hold for division.
$3 \div (1 + 2) \neq (3 \div 1) + (3 \div 2)$

Check It Out

Use the distributive property to find each product.

13. $6 \cdot 98$
14. $3 \cdot 105$
15. $4 \cdot 199$
16. $5 \cdot 214$

Equivalent Expressions

The distributive property can be used to write an **equivalent expression** with two terms. Equivalent expressions are two different ways of writing one expression.

WRITING AN EQUIVALENT EXPRESSION

Write an equivalent expression for $2(3x - 1)$.

$2\,(3x - 1)$ • Distribute the factor to each term inside the parentheses.

$= 2 \cdot 3x - 2 \cdot 1$ • Simplify.

$= 6x - 2$ • Write the equivalent expressions.

$2(3x - 1) = 6x - 2$

Distributing When the Factor Is Negative

The distributive property is applied in the very same way if the factor to be distributed is negative.

DISTRIBUTING WHEN THE FACTOR IS NEGATIVE

Write an equivalent expression for $-3(5x - 1)$.

$-3\,(5x - 1)$ • Distribute the factor to each term inside the parentheses.

$= -3 \cdot 5x - (-3) \cdot 1$ • Simplify.
Remember: $(-3) \cdot 1 = -3$
and $-(-3) = +3$.

$= -15x + 3$ • Write the equivalent expressions.

$-3(5x - 1) = -15x + 3$

Check It Out

Write an equivalent expression.

17. $2(5x + 2)$ 18. $6(3n - 2)$

19. $-1(6y - 4)$ 20. $-2(-4x + 5)$

The Distributive Property with Common Factors

For the expression $10x + 15$, you can use the distributive property to write an equivalent expression. Recognize that each of the two terms has a factor of 5.

Rewrite the expression as $5 \cdot 2x + 5 \cdot 3$. Then write the common factor 5 in front of the parentheses and the remaining factors inside the parentheses: $5(2x + 3)$. You have used the distributive property to *factor out the common factor.*

FACTORING OUT THE COMMON FACTOR

Factor out the common factor from the expression $12n - 30$.

$12n - 30$	• Find a common factor.
$6 \cdot 2n - 6 \cdot 5$	• Rewrite the expression.
$6 \cdot (2n - 5)$	• Use the distributive property.

$12n - 30 = 6 \cdot (2n - 5)$

When you factor, always be sure that you factor out the greatest common factor.

Check It Out

Factor out the greatest common factor in each expression.

21. $7x + 21$
22. $12n - 9$
23. $10c + 30$
24. $10a - 25$

Like Terms

Like terms are terms that contain the same variable with the same exponent. Constant terms are like terms because they do not have any variables. Here are some examples of like terms:

Like Terms	Reason
$3x$ and $4x$	Both contain the same variables.
3 and 11	Both are constant terms.
$2n^2$ and $6n^2$	Both contain the same variable with the same exponent.

Some examples of terms that are not like terms are listed.

Not Like Terms	Reason
$3x$ and $5y$	Variables are different.
$4n$ and 12	One term has a variable; the other is constant.
$2x^2$ and $2x$	The variables are the same, but the exponents are different.

Two terms that are like terms may be combined into one term by adding or subtracting. Consider the expression $3x + 4x$. Notice that the two terms have a common factor, x. Use the distributive property to write $x(3 + 4)$. This simplifies to $7x$, so $3x + 4x = 7x$.

COMBINING LIKE TERMS

Simplify $4y - 8y$.

- Recognize that the variable is a common factor. Rewrite the expression, using the distributive property.

  $y(4 - 8)$

- Simplify.

  $y(-4)$

- Use the commutative property of multiplication.

  $-4y$

6.2 SIMPLIFYING EXPRESSIONS

Check It Out
Combine like terms.
25. $4x + 9x$
26. $10y - 6y$
27. $5n + 4n + n$
28. $3a - 7a$

Simplifying Expressions

Expressions are simplified when all of the like terms have been combined. Terms that are not like terms cannot be combined. In the expression $3x - 5y + 6x$, there are three terms. Two of them are like terms, $3x$ and $6x$, which combine to be $9x$. The expression can be written as $9x - 5y$, which is simplified because the two terms are not like terms.

SIMPLIFYING EXPRESSIONS

Simplify the expression $4(2n - 3) - 10n + 17$.

$4(2n - 3) - 10n + 17$	• Combine like terms if any.
	• Use the distributive property.
$= 4 \cdot 2n - 4 \cdot 3 - 10n + 17$	• Simplify.
$= 8n - 12 - 10n + 17$	• Combine like terms.
$= -2n + 5$	• If remaining terms are not like terms, the expression is simplified.

 Check It Out
Simplify each expression.
29. $7y + 5z - 2y + z$
30. $x + 3(2x - 5)$
31. $10a + 6 - 2(3a + 2)$
32. $2(4n - 3) - (n - 2)$

6·2 EXERCISES

Decide whether each term is a constant term.
1. $8n$ 2. -7

Rewrite each expression and use the commutative property of addition or multiplication.
3. $4 + 6$ 4. $n \cdot 7$ 5. $4x + 5$

Rewrite each expression and use the associative property of addition or multiplication.
6. $4 + (5 + 9)$ 7. $(7 \cdot 5) \cdot 2$
8. $2 \cdot 5n$

Use the distributive property to find each product.
9. $4 \cdot 99$ 10. $6 \cdot 104$

Write an equivalent expression.
11. $3(5x + 4)$ 12. $-5(2n + 6)$
13. $10(3a - 7)$ 14. $-(-4y - 6)$

Factor out the greatest common factor in each expression.
15. $8x + 16$ 16. $12n - 4$
17. $20a - 30$

Combine like terms.
18. $10x - 7x$ 19. $5n + 6n - n$
20. $3a - 8a$

Simplify each expression.
21. $8a + b - 2a - 4b$ 22. $5x + 2(3x - 5) + 2$
23. $-2(-5n - 3) - (n + 2)$
24. Which property is illustrated by $5(2x + 1) = 10x + 5$?
    A. Commutative property of multiplication
    B. Distributive property
    C. Associative property of multiplication
    D. The example does not illustrate a property.
25. Which of the following shows the expression $24x - 36$ with the greatest common factor factored out?
    A. $2(12x - 18)$ B. $3(8x - 12)$
    C. $6(4x - 6)$ D. $12(2x - 3)$

6·3 Evaluating Expressions and Formulas

Evaluating Expressions

Once an expression has been written, you can *evaluate* it for different values of the variable. To evaluate $2x - 1$ for $x = 4$, *substitute* 4 in place of the x: $2(4) - 1$. Use **order of operations** to evaluate: multiply first, then subtract. So $2(4) - 1 = 8 - 1 = 7$.

EVALUATING AN EXPRESSION

Evaluate $2x^2 - \frac{6}{x} + 3$ for $x = 3$.

$2x^2 - \frac{6}{x} + 3$, when $x = 3$ • Substitute numeric value for variable.

$= 2(3^2) - \frac{6}{(3)} + 3$ • Use order of operations to simplify. Simplify within parentheses, then evaluate.

$= 2 \cdot 9 - \frac{6}{3} + 3$ • Multiply and divide, in order from left to right.

$18 - 2 + 3 = 19$ • Add and subtract, in order from left to right.

When $x = 3$, then $2x^2 - \frac{6}{x} + 3 = 19$.

Check It Out

Evaluate each expression for the given value.

1. $6x - 10$, for $x = 4$
2. $4a + 5 + a^2$, for $a = -3$
3. $\frac{n}{4} + 2n - 3$, for $n = 8$
4. $2(y^2 - y - 2) + 2y$, for $y = 3$

Evaluating Formulas

The Formula for Perimeter of a Rectangle

The **perimeter** of a rectangle is the distance around the rectangle. The **formula** $P = 2w + 2l$ can be used to find the perimeter, P, if the width, w, and the length, l, are known.

FINDING THE PERIMETER OF A RECTANGLE

Find the perimeter of a rectangle whose width is 2 ft and length is 4 ft.

$P = 2(2) + 2(4)$ • Substitute values into formula for perimeter of a rectangle $(P = 2w + 2l)$.

$= 4 + 8$ • Simplify, using order of operations.

$= 12$

The perimeter of the rectangle is 12 feet.

Check It Out

Find the perimeter of each rectangle described.

5. $w = 4$ cm, $l = 9$ cm
6. $w = 10$ m, $l = 15$ m
7. $w = 1$ in., $l = 6$ in.
8. $w = 3.5$ ft, $l = 8.5$ ft

The Formula for Distance Traveled

The distance traveled by a person, vehicle, or object depends on its rate and the amount of time. The formula $d = rt$ can be used to find the distance traveled, d, if the rate, r, and the amount of time, t, are known.

FINDING THE DISTANCE TRAVELED

Find the distance traveled by a runner who averages 5 mi/hr for 4 hr.

$d = 5 \times 4$ • Substitute values into the distance formula $(d = rt)$.

$= 20$ mi • Multiply.

The runner ran 20 miles.

6•3 EVALUATING EXPRESSIONS

Check It Out

Find the distance traveled.

9. A person rides 15 mi/hr for 2 hr.
10. A plane flies 700 km/hr for 3 hr.
11. A person drives a car 55 mi/hr for 6 hr.
12. A snail moves 2 ft/hr for 5 hr.

Maglev

Maglev (short for *magnetic levitation*) trains fly above the tracks. Magnetic forces lift and propel the trains. Without the friction of the tracks, the maglevs run at speeds of 150 to 300 mi/hr. Are they the trains of the future? At a speed of 200 mi/hr with no stops, how long would it take to travel the distance between these cities? Round to the nearest quarter of an hour. See Hot Solutions for answers.

235 mi from Boston, MA to New York, NY
440 mi from Los Angeles, CA to San Francisco, CA
750 mi from Mobile, AL to Miami, FL

6·3 EXERCISES

Evaluate each expression for the given value.
1. $5x - 11$, for $x = 6$
2. $3a^2 + 7 - 2a$, for $a = 4$
3. $\frac{n}{6} - 2n + 8$, for $n = -6$
4. $3(2y - 1) - \frac{12}{y} + 6$, for $y = 4$

Use the formula $P = 2w + 2l$ to answer items 5–7.
5. Find the perimeter of a rectangle that is 40 ft long and 15 ft wide.
6. Find the perimeter of the rectangle.

6 cm

15 cm

7. Kelly had a 20-in. by 30-in. enlargement made of a photograph. She wanted to have it framed. Kelly decided that the picture would look better if there was a 2-in. matte all the way around the photo. How many inches of frame would it take to enclose the photo and matte?

Use the formula $d = rt$ to answer items 8–10.
8. Find the distance traveled by a walker who walks at 4 mi/hr for $1\frac{1}{2}$ hr.
9. A race car driver averaged 140 mi/hr. If the driver completed the race in $2\frac{1}{2}$ hr, how many miles was the race?
10. The speed of light is approximately 186,000 mi/sec. About how far does light travel in 5 sec?

6·4 Solving Linear Equations

Additive Inverses

Two terms are **additive inverses** if their sum is 0. Some examples are -3 and 3, $5x$ and $-5x$, and $12y$ and $-12y$. The additive inverse of 7 is -7, because $7 + (-7) = 0$, and the additive inverse of $-8n$ is $8n$, because $-8n + 8n = 0$.

Check It Out
Give the additive inverse of each term.
1. 4
2. $-x$
3. -35
4. $10y$

True or False Equations

The equation $3 + 4 = 7$ represents a true statement. The equation $1 + 4 = 7$ represents a false statement. What about the equation $x + 4 = 7$? You cannot determine whether it is true or false until a value for x is known.

DETERMINE IF THE EQUATION IS TRUE OR FALSE

Determine whether the equation $3x - 2 = 13$ is true or false for $x = 1$, $x = 3$, and $x = 5$.

$3x - 2 = 13$	$3x - 2 = 13$	$3x - 2 = 13$
$3(1) - 2 \stackrel{?}{=} 13$	$3(3) - 2 \stackrel{?}{=} 13$	$3(5) - 2 \stackrel{?}{=} 13$
$3 - 2 \stackrel{?}{=} 13$	$9 - 2 \stackrel{?}{=} 13$	$15 - 2 \stackrel{?}{=} 13$
$1 \stackrel{?}{=} 13$	$7 \stackrel{?}{=} 13$	$13 \stackrel{?}{=} 13$
False	False	True

6·4 SOLVING LINEAR EQUATIONS

Check It Out

Determine whether each equation is true or false for $x = 2$, $x = 5$, and $x = 8$.

5. $6x - 3 = 9$
6. $2x + 3 = 13$
7. $5x - 7 = 18$
8. $3x - 8 = 16$

The Solution of an Equation

If you look back over the past equations, you will notice that each equation had only one value for the variable that made the equation true. This value is called the **solution** of the equation. If you were to try other values for x in the equations, they would all give false statements.

DETERMINING A SOLUTION

Determine whether 6 is the solution of the equation $4x - 5 = 2x + 6$.

$4x - 5 = 2x + 6$ • Substitute possible solution for x.

$4(6) - 5 \overset{?}{=} 2(6) + 6$ • Simplify, using order of operations.

$24 - 5 \overset{?}{=} 12 + 6$

$19 \overset{?}{=} 18$

Since the statement is false, 6 is not the solution.

Check It Out

Determine whether the given value is the solution of the equation.

9. $6; 3x - 5 = 13$
10. $7; 2n + 5 = 3n - 5$
11. $4; 7(y - 2) = 10$
12. $1; 5x + 4 = 12x - 3$

Equivalent Equations

An *equivalent equation* can be obtained from an existing equation in one of four ways.

- Add the same term to both sides of the equation.
- Subtract the same term from both sides.
- Multiply by the same term on both sides.
- Divide by the same term on both sides.

Four equations equivalent to $x = 8$ are shown.

Operation	Equation Equivalent to $x = 8$
Add 4 to both sides.	$x + 4 = 12$
Subtract 4 from both sides.	$x - 4 = 4$
Multiply by 4 on both sides.	$4x = 32$
Divide by 4 on both sides.	$\frac{x}{4} = 2$

Check It Out

Write equations equivalent to $x = 12$.
13. Add 3 to both sides.
14. Subtract 3 from both sides.
15. Multiply by 3 on both sides.
16. Divide by 3 on both sides.

Solving Equations

You can use equivalent equations to *solve* an equation. The solution is obtained when the variable is by itself on one side of the equation. The objective, then, is to use equivalent equations to isolate the variable on one side of the equation.

Consider the equation $x + 7 = 15$. For it to be considered solved, the x has to be on a side by itself. How can you get rid of the $+7$ that is also on that side? Remember that a term and its additive inverse add up to 0. The additive inverse of 7 is -7. To write an equivalent equation, subtract 7 from both sides.

$$x + 7 = 15$$

• Subtract 7 from both sides.

$$x + 7 - 7 = 15 - 7$$

• Simplify.

$$x = 8$$

You can check the solution to be sure it is correct.

$$x + 7 = 15$$

• Substitute the possible solution for *x*.

$$8 + 7 \stackrel{?}{=} 15$$

• Simplify.
• Since this is a true statement,

$$15 \stackrel{?}{=} 15$$

8 is the solution.

To solve the equation $n - 3 = 10$, you have to write an equivalent equation with *n* on a side by itself. Notice that there is a -3 on the same side. Its additive inverse is $+3$, so add 3 to both sides of the equation.

$$n - 3 = 10$$

• Add 3 to both sides.

$$n - 3 + 3 = 10 + 3$$

• Simplify.

$$n = 13$$

Check the solution.

$$n - 3 = 10$$

• Substitute the possible solution for *n*.

$$(13) - 3 \stackrel{?}{=} 10$$

• Simplify.
• Since this is a true statement,

$$10 \stackrel{?}{=} 10$$

13 is the solution.

Check It Out

Solve each equation. Check your solution.

17. $x + 4 = 11$

18. $n - 5 = 8$

19. $y + 8 = 2$

20. $a - 5 = 1$

More Solving Equations

Consider the equation $3x = 15$. Notice that there isn't a term being added to or subtracted from the term with the variable. However, the variable still is not isolated. The variable is being multiplied by 3. To write an equivalent equation with the variable isolated, divide by 3 on both sides.

$3x = 15$ • Divide by 3 on both sides.

$\frac{3x}{3} = \frac{15}{3}$ • Simplify.

$x = 5$

Check the solution.

$3x = 15$ • Substitute the possible solution for x.

$3(5) \overset{?}{=} 15$ • Simplify.

$15 \overset{?}{=} 15$ • Since this is a true statement, 5 is the solution.

To solve the equation $\frac{n}{6} = 3$, notice that the variable is not isolated. The variable is being divided by 6. To write an equivalent equation with the variable isolated, multiply by 6 on both sides.

$\frac{n}{6} = 3$ • Multiply by 6 on both sides.

$\frac{n}{6} \cdot 6 = 3 \cdot 6$ • Simplify.

$n = 18$

Check the solution.

$\frac{n}{6} = 3$ • Substitute the possible solution for n.

$\frac{(18)}{6} \overset{?}{=} 3$ • Simplify.

$3 \overset{?}{=} 3$ • Since this is a true statement, 18 is the solution.

Check It Out

Solve each equation. Check your solution.

21. $6x = 30$

22. $\frac{y}{4} = 5$

23. $9n = -27$

24. $\frac{a}{3} = 10$

Solving Equations Requiring Two Operations

In the equation $2x - 3 = 11$, notice that the variable is being multiplied and has a term being subtracted. Still the objective is to use equivalent equations to isolate the variable. To do this, first isolate the term that contains the variable. Then isolate the variable.

$2x - 3 = 11$ • Add 3 to both sides to isolate the term that contains the variable.

$2x - 3 + 3 = 11 + 3$ • Simplify.

$2x = 14$ • Divide by 2 on both sides to isolate the variable.

$\dfrac{2x}{2} = \dfrac{14}{2}$ • Simplify.

$x = 7$

Check the solution.

$2x - 3 = 11$ • Substitute the possible solution for x.

$2(7) - 3 \overset{?}{=} 11$ • Simplify, using order of operations.

$14 - 3 \overset{?}{=} 11$

$11 \overset{?}{=} 11$ • Since this is a true statement, 7 is the solution.

SOLVING EQUATIONS REQUIRING TWO OPERATIONS

Solve the equation $\frac{n}{3} + 1 = 3$.

$\frac{n}{3} + 1 = 3$ • Add or subtract on both sides to isolate the term containing the variable.

$\frac{n}{3} + 1 - 1 = 3 - 1$ • Simplify.

$\frac{n}{3} = 2$ • Multiply or divide on both sides to isolate the variable.

$\frac{n}{3} \times 3 = 2 \times 3$ • Simplify.

$n = 6$ • Check solution by substituting into original equation.

$\frac{(6)}{3} + 1 \stackrel{?}{=} 3$ • Simplify, using order of operations.

$2 + 1 \stackrel{?}{=} 3$

$3 \stackrel{?}{=} 3$ • If the statement is true, you have the solution.

When $\frac{n}{3} + 1 = 3$, n is 6.

Check It Out

Solve each equation. Check your solution.

25. $4x + 7 = 27$

26. $\frac{y}{5} - 2 = 8$

27. $2n + 11 = 3$

28. $\frac{a}{3} + 7 = 5$

Solving Equations with the Variable on Both Sides

Consider the equation $x - 7 = -2x + 5$. Notice that both sides of the equation have a term with the variable. To solve this equation, you still have to use equivalent equations to isolate the variable.

To isolate the variable, first use the additive inverse of one of the terms that contain the variable to collect these terms on one side of the equation. (Generally, this should be on the side of the equation where the coefficient of the variable is higher—this allows you to work with positive numbers whenever possible.) Then use the additive inverse to collect the constant terms on the other side. Then multiply or divide to isolate the variable.

SOLVING AN EQUATION WITH VARIABLES ON BOTH SIDES

Solve the equation $x - 7 = -2x + 5$.

$x - 7 + 2x = -2x + 5 + 2x$
- Add or subtract on both sides to collect terms with the variable on one side.

$3x - 7 = 5$
- Simplify. Combine like terms.

$3x - 7 + 7 = 5 + 7$
- Add or subtract on both sides to collect constant terms on the side opposite the variable.

$3x = 12$
- Simplify.

$\dfrac{3x}{3} = \dfrac{12}{3}$
- Multiply or divide on both sides to isolate the variable.

$x = 4$
- Simplify.

$4 - 7 \overset{?}{=} -2(4) + 5$
- Check by substituting possible solution into original equation.

$4 - 7 \overset{?}{=} -8 + 5$
- Simplify, using order of operations.

$-3 \overset{?}{=} -3$
- If the statement is true, you substituted the correct solution.

For $x - 7 = -2x + 5$, $x = 4$

6-4 SOLVING LINEAR EQUATIONS

Check It Out

Solve each equation. Check your solution.

29. $8n - 4 = 5n + 8$

30. $12x + 5 = 2x - 15$

Equations Involving the Distributive Property

To solve the equation $3x - 4(2x + 5) = 3(x - 2) + 10$, notice that the terms are not yet ready to be collected on one side of the equation. First you have to use the distributive property.

$3x - 4(2x + 5) = 3(x - 2) + 10$ • Simplify, using distributive property.

$3x - 8x - 20 = 3x - 6 + 10$ • Combine like terms.

$-5x - 20 = 3x + 4$ • Add or subtract on both sides to collect terms with variable on one side.

$-5x - 20 + 5x = 3x + 4 + 5x$ • Combine like terms.

$-20 = 8x + 4$ • Add or subtract on both sides to collect constant terms on the side opposite from the variable.

$-20 - 4 = 8x + 4 - 4$ • Combine like terms.

$-24 = 8x$ • Multiply or divide on both sides to isolate the variable.

$\dfrac{-24}{8} = \dfrac{8x}{8}$ • Simplify.

$-3 = x$ • Substitute the possible solution into the original equation.

$3(-3) - 4[2(-3)+5] \overset{?}{=} 3[(-3)-2] + 10$ • Simplify, using order of operations.
$3(-3) - 4(-6+5) \overset{?}{=} 3(-5) + 10$
$-9 - (-4) \overset{?}{=} -15 + 10$

$-5 \overset{?}{=} -5$ • If the statement is true, you substituted the correct solution.

Check It Out

Solve each equation. Check your solution.

31. $4(n - 2) = 12$

32. $6 - 2(x - 2) = 6(x + 3)$

Solving for a Variable in a Formula

Recall the formula $d = rt$, in which the distance traveled, d, was found by multiplying the average rate, r, by the amount of time traveled, t. Could you solve for t in the formula?

Notice that the r and the t are being multiplied. To write an equivalent equation so that the t is isolated, divide by r on both sides.

$d = rt$ • Divide by r on both sides.

$\dfrac{d}{r} = \dfrac{rt}{r}$ • Simplify.

$\dfrac{d}{r} = t$

You can solve for w in the formula for the perimeter of a rectangle, $P = 2w + 2l$.

$P = 2w + 2\ell$ • To isolate the term that contains w, subtract $2l$ from both sides.

$P - 2\ell = 2w + 2\ell - 2\ell$ • Combine like terms.

$P - 2\ell = 2w$ • To isolate w, divide both sides by 2.

$\dfrac{P - 2\ell}{2} = \dfrac{2w}{2}$ • Simplify.

$\dfrac{P - 2\ell}{2} = w$

Check It Out

Solve for the indicated variable in each formula.

33. $A = lw$, for w

34. $3y + x = 6$, for y

6-4 SOLVING LINEAR EQUATIONS

Three Astronauts and a Cat

Here is a modern version of a problem that first appeared in the year 850.

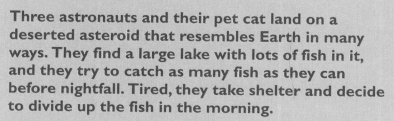

Three astronauts and their pet cat land on a deserted asteroid that resembles Earth in many ways. They find a large lake with lots of fish in it, and they try to catch as many fish as they can before nightfall. Tired, they take shelter and decide to divide up the fish in the morning.

One astronaut wakes up during the night and decides to take her share. She divides the pile of fish into three equal parts, but there is one left over. So she gives it to the cat. She hides her third and puts the rest of the fish back in a pile. Later the second and third astronauts wake up in turn and do exactly the same thing. In the morning they divide the pile of fish that's left into three equal parts. They give the one remaining fish to the cat. What is the smallest number of fish they originally caught? See Hot Solutions for answer.

6·4 EXERCISES

Give the additive inverse of each term.

1. 7 2. $-4x$

Determine whether the given value is the solution of the equation.

3. $6; 3(y - 2) = 12$ 4. $5; 6n - 5 = 3n + 11$

Solve each equation. Check your solution.

5. $x + 7 = 12$
6. $n - 6 = 11$
7. $\frac{y}{8} = 3$
8. $7a = -28$
9. $x + 12 = 7$
10. $n - 11 = 5$
11. $9x = 63$
12. $\frac{a}{6} = -2$
13. $3x + 7 = 25$
14. $\frac{y}{4} - 2 = 5$
15. $2n + 11 = 7$
16. $\frac{a}{3} + 9 = 5$
17. $13x - 5 = 10x + 7$
18. $y + 6 = 3y - 8$
19. $8x + 6 = 3x - 4$
20. $3a + 4 = 4a - 3$
21. $6(2n - 5) = 4n + 2$
22. $9y - 4 - 6y = 2(y + 1) - 5$
23. $8x - 3(x - 1) = 4(x + 2)$
24. $14 - (6x - 5) = 5(2x - 1) - 4x$

Solve for the indicated variable in each formula.

25. $d = rt$, for r 26. $A = lw$, for l
27. $4y - 5x = 12$, for y 28. $8y + 3x = 11$, for y

29. Which of the following equations can be solved by adding 6 to both sides and dividing by 5 on both sides?

A. $5x + 6 = 16$ B. $\frac{x}{5} + 6 = 16$
C. $5x - 6 = 14$ D. $\frac{x}{5} - 6 = 14$

30. Which equation does not have $x = 4$ as its solution?

A. $3x + 5 = 17$ B. $2(x + 2) = 10$
C. $\frac{x}{2} + 5 = 7$ D. $x + 2 = 2x - 2$

6·5 Ratio and Proportion

Ratio

A **ratio** is a comparison of two quantities. If there are 10 boys and 15 girls in a class, the ratio of the number of boys to the number of girls is 10 to 15, which can be expressed as the fraction $\frac{10}{15}$ and reduced to $\frac{2}{3}$. You can write some other ratios.

Comparison	Ratio	As a Fraction
Number of girls to number of boys	15 to 10	$\frac{15}{10} = \frac{3}{2}$
Number of boys to number of students	10 to 25	$\frac{10}{25} = \frac{2}{5}$
Number of students to number of girls	25 to 15	$\frac{25}{15} = \frac{5}{3}$

Check It Out

A coin bank contains four nickels and eight dimes. Write each ratio and reduce to lowest terms.

1. number of nickels to number of dimes
2. number of dimes to number of coins
3. number of coins to number of nickels

Proportions

A **rate** is a ratio that compares a quantity to one unit. Some examples of rates are listed below.

$$\frac{\$8}{1\ hr} \qquad \frac{2\ cans}{\$1} \qquad \frac{12\ in.}{1\ ft} \qquad \frac{35\ mi}{1\ hr} \qquad \frac{24\ mi}{1\ gal}$$

If a car gets $\frac{35\ mi}{1\ gal}$, then the car can get $\frac{70\ mi}{2\ gal}$, $\frac{105\ mi}{3\ gal}$, and so on. The ratio are all equal—they can be reduced to $\frac{35}{1}$.

When two ratios are equal, they form a **proportion.** One way to determine whether two ratios form a proportion is to check their **cross products.** Every proportion has two cross products: the numerator of one ratio multiplied by the denominator of the other ratio. If the cross products are equal, the two ratios form a proportion.

DETERMINING A PROPORTION

Determine whether a proportion is formed.

$\dfrac{6}{9} \overset{?}{\times} \dfrac{45}{60}$ $\dfrac{15}{9} \overset{?}{\times} \dfrac{70}{42}$ • Find the cross products.

$6 \cdot 60 \overset{?}{=} 45 \cdot 9$ $15 \cdot 42 \overset{?}{=} 70 \cdot 9$

$360 \overset{?}{=} 405$ $630 \overset{?}{=} 630$ • If the sides are equal, the ratios are proportional.

$\dfrac{6}{9} \overset{?}{=} \dfrac{45}{60}$ $\dfrac{15}{9} \overset{?}{=} \dfrac{70}{42}$

is not a   is a
proportion.   proportion.

Check It Out

Determine whether a proportion is formed.

4. $\dfrac{4}{7} = \dfrac{12}{21}$ 5. $\dfrac{6}{5} = \dfrac{50}{42}$

Using Proportions to Solve Problems

To use proportions to solve problems, set up two ratios, which relate what you know to what you are solving for.

Suppose that you can buy 5 apples for $2. How much would it cost to buy 17 apples? Let c represent the cost of the 17 apples.

If you express each ratio as $\dfrac{\text{apples}}{\$}$, then one ratio is $\dfrac{5}{2}$ and another is $\dfrac{17}{c}$. The two ratios must be equal.

$\dfrac{5}{2} = \dfrac{17}{c}$

To solve for c, you can use the cross products. Because you have written a proportion, the cross products are equal. To isolate the variable, divide by 5 on both sides and simplify.

$5c = 34$ $\dfrac{5c}{5} = \dfrac{34}{5}$ $c = 6.8$

So 17 apples would cost $6.80.

Check It Out

Use proportions to solve items 6 and 7.

6. A car gets 20 miles per gallon. How many gallons would the car need to travel 90 miles?

7. A worker earns $100 every 8 hours. How much would the worker earn in 28 hours?

Prime Time

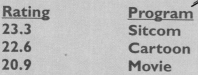

One week a list of the top three rated prime-time TV shows looked like this:

Rating	Program
23.3	Sitcom
22.6	Cartoon
20.9	Movie

Say that *a* equals the total number of homes watching TV. Say that number of homes watching the movie was 39 million (20.9*a*). How many homes watched the sitcom? See Hot Solutions for answer.

6·5 EXERCISES

A basketball team has 15 wins and 5 losses. Write each ratio.

1. number of wins to number of losses

2. number of wins to number of games

3. number of losses to number of games

Determine whether a proportion is formed.

4. $\frac{5}{7} = \frac{8}{11}$

5. $\frac{9}{6} = \frac{15}{10}$

6. $\frac{4}{9} = \frac{11}{24}$

Use a proportion to solve each problem.

7. In a class, the ratio of boys to girls is $\frac{3}{4}$. If there are 12 boys in the class, how many girls are there?

8. An overseas telephone call costs $0.36 per minute. How much would a 5-minute call cost?

9. A map is drawn with a scale of 80 mi to 1 cm. The distance between two cities is 600 mi. How far apart are the two cities on the map?

10. A blueprint of a house is drawn with a scale of 5 ft to 2 cm. On the blueprint, a room is 5 cm long. How long will the actual room be?

6·6 Inequalities

Showing Inequalities

When comparing the numbers 7 and 4, you might say that "7 is greater than 4," or you might also say "4 is less than 7." When two expressions are not equal, or could be equal, you can write an **inequality**. The symbols are shown in the chart.

Symbol	Meaning	Example
>	Is greater than	$7 > 4$
<	Is less than	$4 < 7$
≥	Is greater than or equal to	$x \geq 3$
≤	Is less than or equal to	$-2 \leq x$

The equation $x = 3$ has one solution, 3. The inequality $x > 3$ has an infinite number of solutions: 3.001, 3.2, 4, 15, 197, and 955 are just some of the solutions. Note that 3 is not a solution—3 is not greater than 3. Because you cannot list all of the solutions, you can show them on a number line.

To show all the values that are greater than 3, but not including 3, use an open circle on 3 and shade the number line to the right.

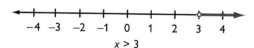

$x > 3$

The inequality $y \leq -2$ also has an infinite number of solutions: -2.01, -2.5, -3, -8, and -54 are just some of the solutions. Note that -2 is also a solution, because -2 is less than *or* equal to -2. On a number line, you want to show all the values that are less than or equal to -2. Because the -2 is to be included, use a closed (filled-in) circle on -2 and shade the number line to the left.

$y \leq -2$

Check It Out

Draw the number line showing the solutions to each inequality.

1. $x \geq 1$
2. $y < -3$
3. $n > -2$
4. $x \leq 4$

Solving Inequalities

Just as you can write equivalent equations, you can write equivalent inequalities. Begin with the inequality $3 > -1$.

$$3 > -1$$

$$-1 \cdot 3 \ ? -1 \cdot -1$$

$$-3 < 1$$

Notice that when the inequality was multiplied or divided by a negative number on both sides, the inequality sign had to be reversed.

Start with 8 > 4. **Perform these operations.**	**Resulting Inequality**
Add 7 to both sides.	$15 > 11$
Subtract 6 from both sides.	$2 > -2$
Multiply by 5 on both sides.	$40 > 20$
Divide by 4 on both sides.	$2 > 1$
Multiply by −3 on both sides.	$-24 < -12$
Divide by −2 on both sides.	$-4 < -2$

To determine the solutions of the inequality $-2x - 3 \geq 3$, use equivalent inequalities to isolate the variable.

$-2x - 3 \geq 3$ • Add or subtract on both sides to isolate the variable term.

$-2x - \underset{\smile}{3 + 3} \geq \underset{\smile}{3 + 3}$ • Combine like terms.

$-2x \geq 6$ • Multiply or divide on both sides to isolate the variable. If you multiply or divide by a negative number, reverse the inequality sign.

$\dfrac{-2x}{-2} \leq \dfrac{6}{-2}$ • Simplify.

$x \leq -3$

Check It Out

Solve each inequality.

5. $x + 9 > 4$
6. $3n \leq -12$
7. $5y + 3 < 18$
8. $-2x + 4 \leq 2$

6·6 EXERCISES

Draw the number line showing the solutions to each inequality.
1. $x < -1$ 2. $y \geq 0$ 3. $n > -3$ 4. $x \leq 5$

Solve each inequality.
5. $x - 4 < 6$ 6. $2y \geq 12$
7. $n + 7 > 4$ 8. $-4a \leq 20$
9. $3x + 4 \geq 16$ 10. $9x - 7 < 11$
11. $-3n + 8 \leq 2$ 12. $10 - y > 6$

13. Which inequality has its solutions represented by the following?

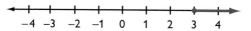

    A. $x < 3$ B. $x \leq 3$ C. $x > 3$ D. $x \geq 3$

14. Which inequality has its solutions represented by the following?

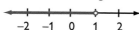

    A. $x \geq 1$ B. $x = 1$ C. $x \leq 1$ D. $x < 1$

15. Which operation(s) would require that the inequality sign be reversed?
    A. Add -3.   B. Subtract -3.
    C. Multiply by -3.   D. Divide by -3.

16. If $x = -1$, is it true that $3(x - 1) \leq 4x$?
17. If $x = 1$, is it true that $2(x - 1) < 0$?

18. Which operation would require that the inequality symbol be reversed?
    A. Multiplication by 4 on both sides
    B. Addition of -4 on both sides
    C. Division by 4 on both sides
    D. Division by -4 on both sides

19. Which of the following statements is false?
    A. $-7 \leq 2$ B. $0 \leq -4$ C. $8 \geq -8$ D. $5 \geq 5$

20. Which of the following inequalities does not have $x < 2$ as its solution?
    A. $-3x < -3$ B. $x + 5 < 7$
    C. $4x - 1 < 7$ D. $-x > -2$

6·7 Graphing on the Coordinate Plane

Axes and Quadrants

When you cross a **horizontal** (left to right) number line with a **vertical** (up and down) number line, the result is a two-dimensional coordinate plane.

The number lines are called **axes.** The horizontal number line is the **x-axis** and the vertical number line is the **y-axis.** The plane is divided into four regions, called **quadrants.** Each quadrant is named by a roman numeral, as shown in the diagram.

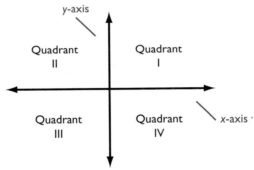

Check It Out

Fill in the blanks.
1. The horizontal number line is called the
   _____.
2. The lower left region of the coordinate plane is called _____.
3. The upper right region of the coordinate plane is called _____.
4. The vertical number line is called the
   _____.

Writing an Ordered Pair

Any location on the coordinate plane can be represented by a **point**. The location of any point is given in relation to where the two axes intersect, called the **origin**.

Two numbers are required to identify the location of a point. The x-coordinate tells you how far to the left or right of the origin the point lies. The y-coordinate tells you how far up or down from the origin the point lies. Together, the x-coordinate and y-coordinate form an **ordered pair**, (x, y).

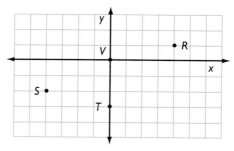

Since point R is 4 units to the right of the origin and 1 unit up, its ordered pair is $(4, 1)$. Point S is 4 units to the left of the origin and 2 units down, so its ordered pair is $(-4, -2)$. Point T is 0 units to the left or right of the origin and 3 units down, so its ordered pair is $(0, -3)$. Point V is 0 units to the left or right of the origin and 0 units up or down. Point V is the origin, and its ordered pair is $(0, 0)$.

 Check It Out

Give the ordered pair for each point.

5. M
6. N
7. P
8. Q

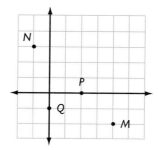

Locating Points on the Coordinate Plane

To locate point $A(3, -4)$ from the origin, move 3 units to the right and 4 units down. Point A lies in Quadrant IV. To locate point $B(-1, 4)$ from the origin, move 1 unit to the left and 4 units up. Point B lies in Quadrant II. Point $C(5, 0)$ is, from the origin, 5 units to the right and 0 units up or down. Point C lies on the x-axis. Point $D(0, -2)$ is, from the origin, 0 units to the left or right and 2 units down. Point D lies on the y-axis.

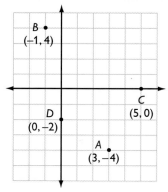

Check It Out

Draw each point on the coordinate plane and tell where it lies.

9. $H(2, -3)$
10. $J(-3, 2)$
11. $K(2, 1)$
12. $L(0, 4)$

The Graph of an Equation with Two Variables

Consider the equation $y = 2x - 1$. Notice that it has two variables, x and y. Point $(3, 5)$ is a solution of this equation. If you substitute 3 for x and 5 for y (in the ordered pair, 3 is the x-coordinate and 5 is the y-coordinate), the true statement $5 = 5$ is obtained. Point $(2, 4)$ is not a solution of the equation. Substituting 2 for x and 4 for y results in the false statement $4 = 3$.

You can generate ordered pairs that are solutions.

Choose a value for x	Substitute the value into the equation $y = 2x - 1$	Solve for y	Ordered Pair
0	$y = 2(0) - 1$	−1	(0, −1)
1	$y = 2(1) - 1$	1	(1, 1)
3	$y = 2(3) - 1$	5	(3, 5)
−1	$y = 2(-1) - 1$	−3	(−1, −3)

If you locate the points on a coordinate plane, you will notice that they all lie along a straight line.

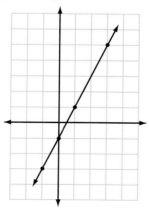

The coordinates of any point on the line will result in a true statement if they are substituted into the equation.

GRAPHING THE EQUATION OF A LINE

Graph the equation $y = \frac{1}{2}x - 1$.

- Choose five values for x.

  Since the value of x is to be multiplied by $\frac{1}{2}$, choose values that are multiples of 2, such as $-2, 0, 2, 4,$ and 6.

- Calculate the corresponding values for y.

  When $x = -2$, $y = \frac{1}{2}(-2) - 1 = -2$.
  When $x = 0$, $y = \frac{1}{2}(0) - 1 = -1$.
  When $x = 2$, $y = \frac{1}{2}(2) - 1 = 0$.
  When $x = 4$, $y = \frac{1}{2}(4) - 1 = 1$.
  When $x = 6$, $y = \frac{1}{2}(6) - 1 = 2$.

- Write the five solutions as ordered pairs (x, y).

  $(-2, -2), (0, -1), (2, 0), (4, 1),$ and $(6, 2)$

- Locate the points on a coordinate plane and draw the line.

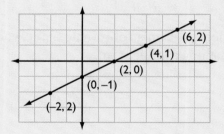

Check It Out

Find five solutions of each equation. Graph each line.

13. $y = 3x - 1$
14. $y = 2x - 1$
15. $y = \frac{1}{2}x - 2$
16. $y = -2x + 1$

More Graphing Equations with Two Variables

How could solutions of the equation $2x - 4y = 8$ be determined? You could use equivalent equations to isolate the y on one side of the equation.

$2x - 4y = 8$	• Add or subtract on both sides to isolate the y term.
$2x - 4y - 2x = 8 - 2x$	• Combine like terms. (Use commutative property to change order of terms.)
$-4y = -2x + 8$	• Multiply or divide on both sides to isolate y.
$\dfrac{-4y}{-4} = \dfrac{-2x + 8}{-4}$	• Simplify.
$y = \dfrac{-2x}{-4} + \dfrac{8}{-4}$	
$y = \dfrac{1}{2}x - 2$	

Now you can find five solutions and graph the line.

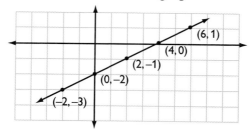

 Check It Out

Graph each line.
17. $2x + y = 3$
18. $x + 2y = 6$
19. $4x - 2y = 4$
20. $2x - 5y = 10$

Horizontal and Vertical Lines

Choose several points that lie on a horizontal line.

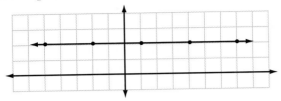

Notice that any point that lies on the line has a y-coordinate of 2. The equation of this line is $y = 2$.

Choose several points that lie on a vertical line.

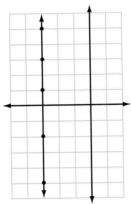

Notice that any point that lies on the line has an x-coordinate of -3. The equation of this line is $x = -3$.

Check It Out
Graph each line.
21. $x = 2$
22. $y = -4$
23. $x = -3$
24. $y = 1$

6·7 EXERCISES

Fill in the blanks.
1. The vertical number line is called the _____.
2. The upper left region of the coordinate plane is called

_____.
3. The lower right region of the coordinate plane is called

_____.

Give the ordered pair for each point.
4. *A*
5. *B*
6. *C*
7. *D*

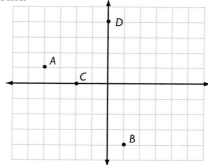

Locate each point on the coordinate plane and tell where it lies.
8. $H(-3, 4)$
9. $J(0, -1)$
10. $K(3, 1)$
11. $L(-2, 0)$

Find five solutions of each equation. Graph each line.
12. $y = 2x - 1$
13. $y = -3x + 2$
14. $y = \frac{1}{2}x - 2$

Graph each line.
15. $2x - y = 2$
16. $x - 3y = 3$
17. $4x + y = 4$
18. $5x + 3y = 15$
19. $x = -2$
20. $y = 5$

6·8 Slope and Intercept

Slope

One characteristic of a line is its **slope**. Slope is a measure of a line's slant. To describe the way a line slants, you need to observe how the coordinates on the line change as you move right. Choose two points along the line. The run is the difference in the x-coordinates. The rise is the difference in the y-coordinates.

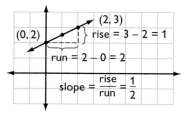

The slope, then, is given by the ratio of rise (vertical movement) to run (horizontal movement).

$$\text{Slope} = \frac{\text{Rise}}{\text{Run}}$$

Notice that for line a, the rise between the two marked points is 10 units and the run is 4 units. The slope of the line, then, is $\frac{10}{4} = \frac{5}{2}$. For line b, the rise is -3 and the run is 6, so the slope of the line is $-\frac{3}{6} = -\frac{1}{2}$.

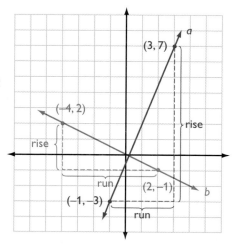

The slope along a straight line is always the same. For line a, regardless of the two points chosen, the slope will always simplify to be $\frac{5}{2}$.

Check It Out
Determine the slope of each line.

1.

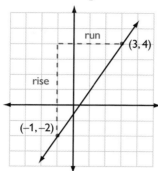

2.

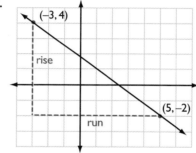

Calculating the Slope of a Line

You can calculate the slope of a line just from knowing
two points on the line. The rise is the difference of the
y-coordinates and the run is the difference of the
x-coordinates. For the line that passes through the points
$(1, -2)$ and $(4, 5)$, the slope can be calculated as shown. The
variable m is used to represent slope.

$$m = \frac{\text{rise}}{\text{run}} = \frac{5 - (-2)}{4 - 1} = \frac{7}{3}$$

The slope could also have been calculated another way.

$$m = \frac{\text{rise}}{\text{run}} = \frac{-2 - 5}{1 - 4} = \frac{-7}{-3} = \frac{7}{3}$$

The order in which you subtract the coordinates does not
matter, as long as you find both differences in the same order.

CALCULATING THE SLOPE OF A LINE

Find the slope of the line that contains the points
$(-2, 3)$ and $(4, -1)$.

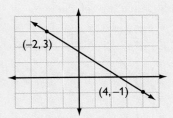

$m = \dfrac{-1-3}{4-(-2)}$ or $m = \dfrac{3-(-1)}{-2-4}$ •Use the definition $m = \dfrac{\text{rise}}{\text{run}} = \dfrac{\text{difference of y-coordinates}}{\text{difference of }x\text{-coordinates}}$ to find the slope.

$m = \dfrac{-4}{6}$ or $m = \dfrac{4}{-6}$ •Simplify.

$m = \dfrac{-2}{3}$ or $m = \dfrac{2}{-3}$

The slope is $-\dfrac{2}{3}$.

Check It Out

Find the slope of the line that contains the given points.

3. $(-2, 7)$ and $(4, 1)$

4. $(-1, -2)$ and $(3, 4)$

5. $(0, 3)$ and $(6, 0)$

6. $(-1, -3)$ and $(1, 5)$

Slopes of Horizontal and Vertical Lines

Choose two points on a horizontal line, $(-1, 2)$ and $(3, 2)$.
Calculate the slope of the line.

$$m = \frac{\text{rise}}{\text{run}} = \frac{2 - 2}{3 - (-1)} = \frac{0}{4} = 0$$

A horizontal line has no rise; its slope is 0.

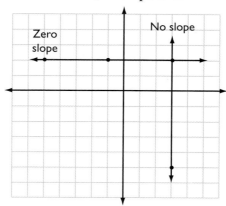

Choose two points on a vertical line, $(3, 2)$ and $(3, -5)$.
Calculate the slope of the line.

$$m = \frac{\text{rise}}{\text{run}} = \frac{-5 - 2}{3 - 3} = \frac{-7}{0}, \text{ which is undefined.}$$

A vertical line has no run; it has *no slope*.

Check It Out

Find the slope of the line that contains the given points.

7. $(-2, 3)$ and $(4, 3)$
8. $(1, -2)$ and $(1, 5)$
9. $(-4, 0)$ and $(-4, 6)$
10. $(4, -2)$ and $(-1, -2)$

6·8 SLOPE AND INTERCEPT

The y-Intercept

A second characteristic of a line, after the slope, is the **y-intercept.** The y-intercept is the location along the y-axis where the line crosses, or intercepts, the y-axis.

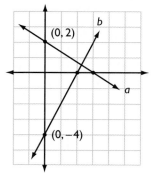

The y-intercept of line *a* is 2, and the y-intercept of line *b* is −4.

 Check It Out

Identify the y-intercept of each line.

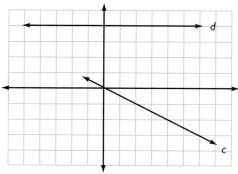

11. *c*

12. *d*

Using the Slope and y-Intercept to Graph a Line

A line can be graphed if the slope and the *y*-intercept are known. First you locate the *y*-intercept on the *y*-axis. Then you use the rise and the run of the slope to locate a second point on the line. Connect the two points to plot your line.

USING THE SLOPE AND y-INTERCEPT TO GRAPH A LINE

Graph the line with slope 2 and *y*-intercept 2.

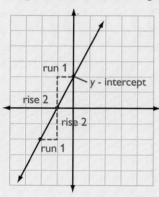

run 1
rise 2
y - intercept
rise 2
run 1

• Locate the *y*-intercept.

• Use the slope to locate other points on the line. If the slope is a whole number *m*, remember $m = \frac{m}{1}$, so rise is *m* and run is 1.

• Draw a line through the points.

Check It Out

Graph each line.

13. Slope = $\frac{1}{3}$; *y*-intercept at -1.

14. Slope = $\frac{-2}{5}$; *y*-intercept at 2.

15. Slope = 2; *y*-intercept at -1.

16. Slope = -3; *y*-intercept at 0.

Slope-Intercept Form

The equation $y = mx + b$ is in the *slope-intercept form* for the equation of a line. When an equation is in this form, the slope of the line is given by m and the y-intercept is located at b. The graph of the equation $y = \frac{2}{3}x - 4$ is a line that has a slope of $\frac{2}{3}$ and a y-intercept at -4. The graph is shown.

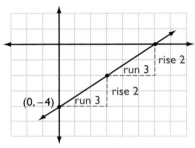

Check It Out

Determine the slope and the y-intercept from the equation of each line.

17. $y = -3x + 2$
18. $y = \frac{1}{4}x - 2$
19. $y = \frac{-2}{3}x$
20. $y = 6x - 5$

Writing Equations in Slope-Intercept Form

To write the equation $2x - y = 3$ in slope-intercept form, isolate the y on one side of the equation. You can use equivalent equations to isolate the y.

$2x - y = 3$	• Add or subtract to isolate the term with the y.
$2x - y - 2x = 3 - 2x$	• Combine like terms. Use commutative property to reorder.
$-y = -2x + 3$	• Multiply or divide to isolate the y.
$\dfrac{-y}{-1} = \dfrac{-2x + 3}{-1}$	• Simplify.
$y = \dfrac{-2}{-1}x + \dfrac{3}{-1}$	
$y = 2x - 3$	

In slope-intercept form, the equation $2x - y = 3$ is $y = 2x - 3$. The slope is 2, and the y-intercept is located at -3. The graph of the line is shown.

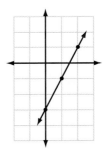

Check It Out

Write each equation in slope-intercept form. Graph the line.

21. $x + 2y = 4$
22. $2x - 3y = 6$
23. $4x - 2y = 8$
24. $5x + y = 6$

Slope-Intercept Form and Horizontal and Vertical Lines

The equation of a horizontal line is in the form $y =$ (number). In the graph below, the horizontal line has the equation $y = 2$. Is this equation in slope-intercept form? Yes, because the equation could be written $y = 0x + 2$. The y is isolated on one side of the equation. The slope, remember, is 0, and the y-intercept is 2.

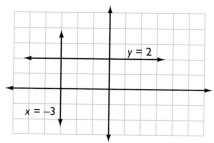

The equation of a vertical line is in the form $x =$ (number). In the graph above, the vertical line has the equation $x = -3$. Is this equation in slope-intercept form? No, because there isn't a y isolated on one side of the equation. A vertical line, remember, has no slope, nor does it have a y-intercept.

Check It Out

Give the slope and y-intercept of each line. Graph the line.

25. $y = -2$ 26. $x = 3$

27. $y = 4$ 28. $x = -1$

Writing the Equation of a Line

If you know the slope and the y-intercept of a line, you can write the equation of the line. If a line has a slope of 1 and a y-intercept of -3, substitute 1 for m and -3 for b into the slope-intercept form for the equation of a line. The equation of the line is $y = x - 3$.

WRITING THE EQUATION OF A LINE

Write the equation of the line.

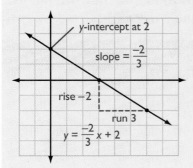

- Identify the *y*-intercept (*b*).

- Find the slope $(m = \dfrac{\text{rise}}{\text{run}})$.

- Substitute the *y*-intercept and the slope into the slope-intercept form ($y = mx + b$).

Check It Out

Write the equation of each line.

29. Slope = 2; *y*-intercept at −4.
30. Slope = $\frac{-2}{3}$; *y*-intercept at 2.
31.

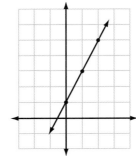

32.

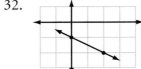

6•8 SLOPE AND INTERCEPT

Writing the Equation of a Line from Two Points

To find the equation of a line when you know only two points, first find the slope and then find the *y*-intercept.

Write the equation of the line that contains the points $(2, 3)$ and $(4, -2)$.

$m = \dfrac{3 - (-2)}{2 - 4} = \dfrac{5}{-2} = \dfrac{-5}{2}$

• Calculate slope, using $m = \dfrac{\text{rise}}{\text{run}}$.

slope $= m = \dfrac{-5}{2}$

• Substitute slope for *m* in slope-intercept form $(y = mx + b)$.

$y = \dfrac{-5}{2}x + b$

• Solve for *b*. Remember that the two given points must be solutions of the equation. Substitute *y*-coordinates for *y* and *x*-coordinates for *x*.

$3 = \dfrac{-5}{2}(2) + b$ or $-2 = \dfrac{-5}{2}(4) + b$

• Simplify.

$3 = -5 + b$ or $-2 = -10 + b$

• Add or subtract to isolate *b*.

• Combine like terms.

$3 + 5 = -5 + b + 5$ or $-2 + 10 = -10 + b + 10$

$8 = b$ or $8 = b$

$y = \dfrac{-5}{2}x + 8$

• Substitute the values you found for *m* and *b* into the slope-intercept form.

Check It Out

Write the equation of the line with the given points.

33. $(-2, 7)$ and $(3, 2)$

34. $(6, 2)$ and $(-3, 0)$

6·8 EXERCISES

Determine the slope of each line for items 1–6.

1.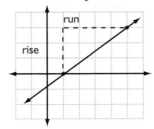

2.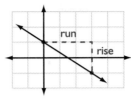

3. Contains $(-1, 3)$ and $(-1, 0)$.
4. Contains $(-4, 2)$ and $(4, -2)$.
5. Contains $(0, -3)$ and $(2, 6)$.
6. Contains $(5, -2)$ and $(5, 4)$.

Graph each line.
7. Slope $= \frac{-1}{3}$; y-intercept at 3.   8. Slope $= 3$; y-intercept at -4.
9. Slope $= -1$; y-intercept at 2.   10. Slope $= 0$; y-intercept at 1.

Determine the slope and the y-intercept from the equation of each line.
11. $y = -2x - 3$
12. $y = \frac{3}{4}x + 2$
13. $y = x + 1$
14. $y = -4$
15. $x = 3$

Write each equation in slope-intercept form. Graph the line.
16. $2x + y = 2$ 17. $x + 4y = 4$
18. $3x - 2y = 6$ 19. $x - y = 3$

Write the equation of each line.
20. Slope $= 2$; y-intercept at -5.
21. Slope $= \frac{-1}{2}$; y-intercept at 3.

Write the equation of the line that contains the given points.
22. $(-4, 5)$ and $(6, 0)$
23. $(4, -3)$ and $(-1, 7)$
24. $(-2, -6)$ and $(3, -1)$
25. $(-2, -8)$ and $(1, 4)$

6·8 EXERCISES

Problem Set

Write an equation for each sentence. **6•1**
1. If 5 is subtracted from the product of 3 and a number, the result is 7 more than the number.
2. 4 times the sum of a number and 3 is 6 less than twice the number.

Factor out the greatest common factor in each expression. **6•2**
3. $5x + 35$ 4. $8n - 6$

Simplify each expression. **6•2**
5. $8a - 3b - a + 7b$
6. $4(2n - 1) - (2n + 3)$

7. Find the distance traveled by an in-line skater who skates at 10 mi/hr for $1\frac{1}{2}$ hr. Use the formula $d = rt$. **6•3**

Solve each equation. Check your solution. **6•4**
8. $x + 7 = 12$ 9. $\frac{y}{3} = -5$
10. $4x - 5 = 19$ 11. $\frac{y}{2} - 4 = 2$
12. $6n - 7 = 2n + 9$ 13. $y - 3 = 7y + 9$
14. $7(n - 2) = 2n + 6$ 15. $10x - 3(x - 1) = 4(x + 3)$

Use a proportion to solve each problem. **6•5**
16. In a class, the ratio of boys to girls is $\frac{3}{2}$. If there are 10 girls in the class, how many boys are there?
17. A map is drawn using a scale of 60 mi to 1 cm. On the map, the two cities are 5.5 cm apart. What is the actual distance between the two cities?

Solve each inequality. Graph the solution. **6•6**

18. $x + 7 \leq 5$ 19. $2x + 6 > 2$

Locate each point on the coordinate plane and tell where it lies.

20. $A(3, 1)$ 21. $B(-2, 0)$ 22. $C(-1, -4)$ 23. $D(0, 3)$

24. Find the slope of the line that contains the points $(5, -2)$ and $(-1, 2)$.

Determine the slope and the y-intercept from the equation of each line. Graph the line.

25. $y = \frac{1}{2}x + 2$ 26. $y = -1$ 27. $x + 2y = 4$ 28. $4x + 2y = 0$

Write the equation of the line that contains the given points.

29. $(-2, 4)$ and $(7, -5)$

30. $(6, -5)$ and $(-3, 1)$

**WRITE DEFINITIONS FOR THE
FOLLOWING WORDS.**

hot **words**

additive inverse
   6•4
associative
   property **6•2**
axes **6•7**
commutative
   property **6•2**
cross product **6•5**
difference **6•1**
distributive
   property **6•2**
equation **6•1**

equivalent **6•1**
equivalent
   expression **6•2**
expression **6•1**
formula **6•3**
horizontal **6•7**
inequality **6•6**
like terms **6•2**
order of
   operations **6•3**
ordered pair **6•7**
origin **6•7**
perimeter **6•3**
point **6•7**
product **6•1**

proportion **6•5**
quadrant **6•7**
quotient **6•1**
rate **6•5**
ratio **6•5**
slope **6•8**
solution **6•4**
sum **6•1**
term **6•1**
variable **6•1**
vertical **6•7**
x-axis **6•7**
y-axis **6•7**
y-intercept **6•8**

WHAT HAVE YOU LEARNED?

Geometry

What do you already know?

You can use the problems and the list of words that follow to see what you already know about this chapter. The answers to the problems are in Hot Solutions at the back of the book, and the definitions of the words are in Hot Words at the front of the book. You can find out more about a particular problem or word by referring to the boldfaced topic number (for example, **7•2**).

Problem Set

1. Find $m\angle ABD$. **7•1**
2. Name two rays that begin at point A. **7•1**
3. What is the sum of the angles in quadrilateral $ABCD$? **7•2**
4. If $ABCD$ is a rectangle, what is $m\angle DBC$? **7•2**

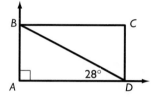

5. What is the measure of each angle in a regular pentagon? **7•2**
6. Which of the letters below have more than one line of symmetry? H E X A G O N **7•3**
7. What is the length of each side of a square that has a perimeter of 56 cm? **7•4**
8. Find the perimeter of a rectangle with a length of 12 m and a width of 7 m. **7•4**
9. What is the area of a triangle with a base of 15 ft and a height of 8 ft? **7•5**
10. The bases of a trapezoid measure 6 in. and 10 in. Its height is 7 in. Find the area of the trapezoid. **7•5**
11. Find the surface area of a rectangular prism that has sides measuring 1 cm, 1 cm, and 17 cm. **7•6**
12. A cylinder with a radius of 10 m is 10 m high. Find the surface area of the cylinder. Use $\pi = 3.14$. **7•6**
13. Find the volume of a cube whose sides measure 5 in. **7•7**
14. The triangular faces of a triangular prism each have an area of 10 cm^2. The prism is 8 cm long. What is its volume? **7•7**
15. The base of a rectangular pyramid measures 6 cm by 12 cm. The pyramid has a height of 20 cm. What is the volume of the pyramid? **7•7**

16. Find the measure of arc $\overparen{AD}$. **7•8**   120°
17. What is the area of circle B in terms of π? **7•8**
    16π m^2

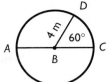

18. Find the length of $\overline{QS}$. **7•9**   5 in.
19. Find the length of $\overline{QR}$. **7•9**   12 in.
20. What is the value of tangent
    $\angle PSQ$? **7•10**   0.75

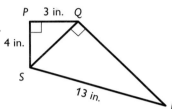

CHAPTER 7

hot **words**

7·1 Naming and Classifying Angles and Triangles

Points, Lines, and Rays

In the world of math, it is sometimes necessary to refer to a specific **point** in space. Simply draw a small dot with a pencil tip to represent a point. A point has no size; its only function is to show position.

Every point needs a name, so to name a point, we use a single capital letter.

·M

Point M

If you draw two points on a sheet of paper, a **line** can be used to connect them. Imagine this line as being perfectly straight and continuing without end in opposite directions. It has no thickness.

Lines need names just like points do, so that we can refer to them easily. To name a line, pick any two points on the line.

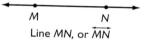

Line MN, or $\overleftrightarrow{MN}$

Because the length of any line is infinite, we sometimes use parts of a line. A **ray** is part of a line that extends without end in one direction. In $\overrightarrow{MN}$, which is read as "ray MN," M is the endpoint. The second point that is used to name the ray can be any point other than the endpoint. You could also name this ray $\overrightarrow{MO}$.

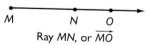

Ray MN, or $\overrightarrow{MO}$

✔ Check It Out

Look at the line below.

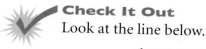

1. Name the line in two different ways.
2. What is the endpoint of $\overrightarrow{KL}$?

Naming Angles

Imagine two different rays with the same endpoint. Together they form what is called an **angle.** The point they have in common is called the **vertex** of the angle. The rays form the sides of the angle.

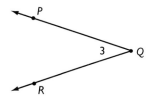

The angle above is made up of $\overrightarrow{QP}$ and $\overrightarrow{QR}$. Q is the common endpoint of the two rays. Point Q is the vertex of the angle. Instead of writing the word *angle,* you can use the symbol for an angle, which is $\angle$.

There are several ways to name an angle. You can name it using the three letters of the points that make up the two rays with the vertex as the middle letter ($\angle PQR$, or $\angle RQP$). You can also use just the letter of the vertex to name the angle ($\angle Q$). Sometimes you might want to name an angle with a number ($\angle 3$).

When more than one angle is formed at a vertex, you use three letters to name each of the angles. Because G is the vertex of three different angles, each angle needs three letters to name it: $\angle DGF$; $\angle DGE$; $\angle EGF$.

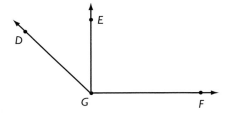

Check It Out
Look at the angles formed by the rays below.

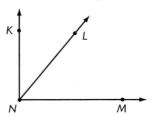

3. Name the vertex.
4. Name all the angles.

Measuring Angles

You measure an angle in **degrees,** using a *protractor* (p. 429).
The number of degrees in an angle will be greater than 0 and
less than or equal to 180.

MEASURING WITH A PROTRACTOR

Measure ∠*ABC*.

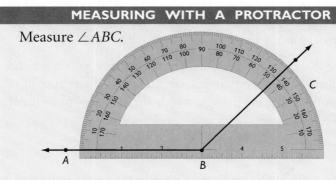

- Place the center point of the protractor on the vertex
  of the angle. Align the 0° line on the protractor with
  one side of the angle.
- Read the number of degrees on the scale where it
  intersects the second side of the angle.

m∠*ABC* = 135°

Check It Out

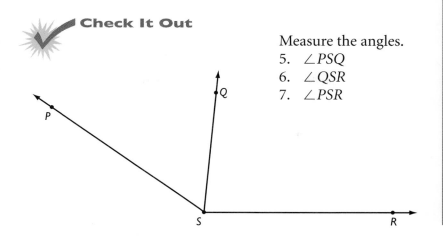

Measure the angles.
5. ∠*PSQ*
6. ∠*QSR*
7. ∠*PSR*

Classifying Angles

You can classify angles by their measures.

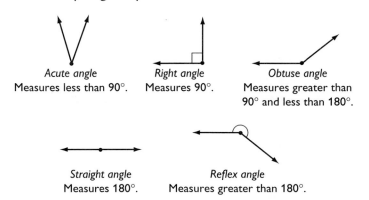

Acute angle
Measures less than 90°.

Right angle
Measures 90°.

Obtuse angle
Measures greater than
90° and less than 180°.

Straight angle
Measures 180°.

Reflex angle
Measures greater than 180°.

Angles that share a side are called *adjacent angles*. You can add measures if the angles are adjacent.

$m\angle KNL = 25°$
$m\angle LNM = 65°$
$m\angle KNM = 25° + 65° = 90°$
Because the sum is 90°,
you know that ∠*KNM* is
a **right angle.**

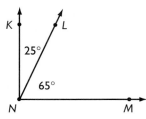

Check It Out
Use a protractor to measure and classify each angle.
8. ∠*SQR*
9. ∠*PQR*
10. ∠*PQS*

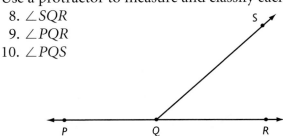

Triangles

Triangles are **polygons** that have three sides, three vertices, and three angles.

You name a triangle with the three vertices in any order. △*ABC* is read "triangle *ABC*."

Classifying Triangles

Like angles, triangles are classified by their angle measures. They are also classified by the number of **congruent** sides, which are sides with equal length.

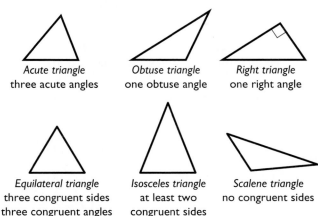

Acute triangle
three acute angles

Obtuse triangle
one obtuse angle

Right triangle
one right angle

Equilateral triangle
three congruent sides
three congruent angles

Isosceles triangle
at least two
congruent sides
at least two
congruent angles

Scalene triangle
no congruent sides

The sum of the measures of the three angles in a triangle is always 180°.

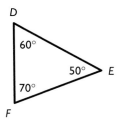

In $\triangle DEF$, $m\angle D = 60°$, $m\angle E = 50°$, and $m\angle F = 70°$.

$60° + 50° + 70° = 180°$

So the sum of the angles of $\triangle DEF$ is 180°.

FINDING THE MEASURE OF THE UNKNOWN ANGLE IN A TRIANGLE

$\angle S$ is a right angle, so its measure is 90°. The measure of $\angle T$ is 35°. Find the measure of $\angle U$.

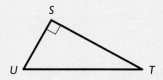

$90° + 35° = 125°$	• Add the two known angles.
$180° - 125° = 55°$	• Subtract the sum from 180°.
$\angle U = 55°$	• The difference is the measure of the third angle.

Check It Out

Find the measure of the third angle of each triangle.

11.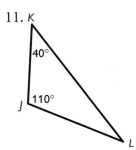

12.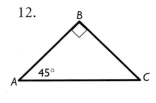

The Triangle Inequality

The length of the third side of a triangle is always less than the sum of the other two sides and greater than their difference. So $(a + b) > c > (a - b)$.

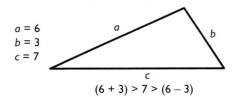

$$a = 6$$
$$b = 3$$
$$c = 7$$

$$(6 + 3) > 7 > (6 - 3)$$

 Check It Out

Which of the following cannot be the lengths of the sides of a triangle?

13. A. 3 ft, 4 ft, 5 ft
    B. 12 m, 2 m, 11 m
    C. 6 in., 10 in., 3 in.
    D. 5 cm, 15 cm, 10 cm
14. A. 7 m, 1 m, 9 m
    B. 9 ft, 9 ft, 9 ft
    C. 12 cm, 12 cm, 3 cm
    D. 6 in., 20 in., 16 in.

7·1 EXERCISES

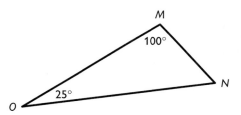

Use the figures to answer the questions.
1. Give six names for the line that passes through point *P.*
2. Name four rays that begin at point *Q.*
3. Name the right angle.
4. Find *m∠PQT.*

5. Find *m∠MNO.*
6. Is △*MNO* an acute, obtuse, or right triangle?
7. Is △*MNO* a scalene, isosceles, or equilateral triangle?

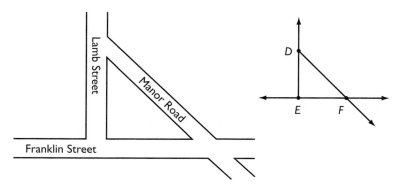

8. Can a triangle have sides measuring 4 ft, 9 ft, and 9 ft? Explain.

9. What line represents Franklin Street?
10. Two streets intersect at a right angle. Name the two streets and the angle.

7·2 Naming and Classifying Polygons and Polyhedrons

Quadrilaterals

You may have noticed that there is a wide variety of four-sided figures, or **quadrilaterals,** to work with in geometry. All quadrilaterals have four sides and four angles. The sum of the angles of a quadrilateral is 360°. There are also many different types of quadrilaterals, which are classified by their sides and angles.

To name a quadrilateral, list the four vertices, either clockwise or counterclockwise.

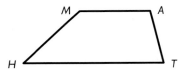

Angles of a Quadrilateral

The sum of the angles of a quadrilateral is 360°.
If you know the measures of three angles in a quadrilateral, you can find the measure of the fourth angle.

FINDING THE MEASURE OF THE UNKNOWN ANGLE IN A QUADRILATERAL

Find $m\angle P$ in quadrilateral PQRS.

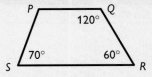

- Add the measures of the three known angles.

$$120° + 60° + 70° = 250°$$

- Subtract the sum from 360°.

$$360° - 250° = 110°$$

- The difference is the measure of the fourth angle.

$$m\angle P = 110°$$

Check It Out

1. Name the quadrilateral in at least two ways.
2. What is the sum of the angles in this quadrilateral?
3. Find $m\angle C$.

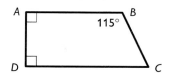

Types of Quadrilaterals

A rectangle is a quadrilateral with four right angles. *EFGH* is one rectangle. Its length is 6 cm and its width is 4 cm.

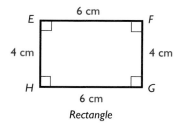

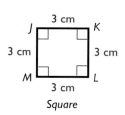

Opposite sides of a rectangle are equal. If all four sides of the rectangle are equal, the rectangle is called a *square.* A square is a **regular shape** because all of its sides are of equal length and all of the interior angles are of equal measure. Some rectangles may be squares, but *all* squares are rectangles. So *JKLM* is a square and a rectangle.

A **parallelogram** is a quadrilateral with opposite sides that are **parallel.** In a parallelogram, opposite sides are equal, and **opposite angles** are equal. *NOPQ* is a parallelogram.

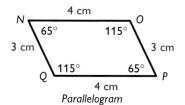

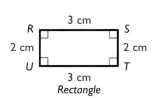

Some parallelograms may be rectangles, but *all* rectangles are parallelograms. Therefore squares are also parallelograms. If all four sides of a parallelogram are the same length, the parallelogram is called a **rhombus.** *WXYZ* is a rhombus.

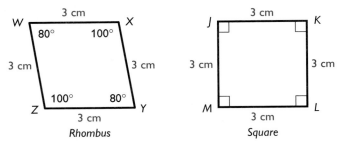

Rhombus   Square

Every square is a rhombus, although not every rhombus is a square, because a square also has equal angles.

In a **trapezoid,** two sides are parallel and two are not. A trapezoid is a quadrilateral, but it is not a parallelogram. *SNAP* is a trapezoid.

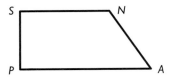

 Check It Out

4. Is quadrilateral *STUV* a rectangle? a parallelogram? a square? a rhombus? a trapezoid?

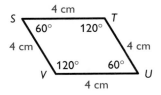

5. Is a square a rhombus? Why or why not?

Polygons

A polygon is a closed figure that has three or more sides. Each side is a **line segment,** and the sides meet only at the endpoints, or vertices.
This figure is a polygon.   These figures are not polygons.

A rectangle, a square, a parallelogram, a rhombus, a trapezoid, and a triangle are all polygons.

There are some aspects of polygons that are always true. For example, a polygon of *n* sides has *n* angles and *n* vertices. A polygon with three sides has three angles and three vertices. A polygon with eight sides has eight angles and eight vertices, and so on.

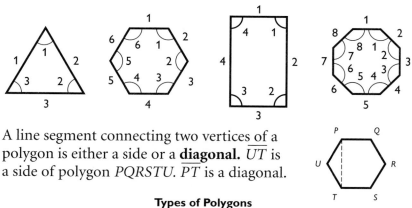

A line segment connecting two vertices of a polygon is either a side or a **diagonal.** $\overline{UT}$ is a side of polygon *PQRSTU*. $\overline{PT}$ is a diagonal.

Types of Polygons

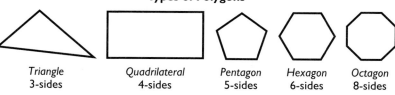

Triangle	Quadrilateral	Pentagon	Hexagon	Octagon
3-sides	4-sides	5-sides	6-sides	8-sides

A seven-sided polygon is called a **heptagon**, a nine-sided polygon is called a **nonagon**, and a ten-sided polygon is called a **decagon.**

Check It Out

State whether or not the figure is a polygon. If it is a polygon, classify it according to the number of sides it has.

6.    7.   8.

A Tangram Zoo

A *tangram* is an ancient Chinese puzzle. The seven tangram pieces fit together to form one large square. To make your own set of tangram pieces, you can cut the shapes, as shown, from a piece of heavy paper or cardboard. You will need a ruler and a protractor, so you can draw right angles and measure the sides as necessary. The areas of the two large triangles are each one-quarter the area of the large square. The area of the small square is one-eighth the area of the large square. And the two small triangles each have one-half the area of the small square.

See if you can use all seven tangram pieces to make this tangram animal. Then see what other tangram pictures you can make.

Angles of a Polygon

You know that the sum of the angles of a triangle is 180° and that the sum of the angles of a quadrilateral is 360°. The sum of the angles of *any* polygon totals at least 180° (triangle). Each additional side adds 180° to the measure of the first three angles. To see why, look at a **pentagon.**

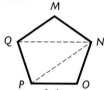

Drawing diagonals $\overline{QN}$ and $\overline{PN}$ shows that the sum of the angles of a pentagon is the sum of the angles in three triangles.

$$3 \times 180° = 540°$$

So the sum of the angles of a pentagon is 540°.

You can use the formula $(n - 2) \times 180°$ to find the sum of the angles of a polygon. Just let n equal the number of sides of a polygon. The answer you get is the sum of the measures of all the angles of the polygon.

FINDING THE SUM OF THE ANGLES OF A POLYGON

$(n - 2) \times 180°$ = sum of angles of a polygon with n sides

Find the sum of the angles of a **hexagon.**

Think: A hexagon has 6 sides. Subtract 2. Then multiply the difference by 180.

- Use the formula: $(6 - 2) \times 180° = 4 \times 180° = 720°$

So the sum of the angles of a hexagon is 720°.

As you know, a regular polygon has equal sides and equal angles. To find the measure of each angle of a regular polygon, you can use what you know about finding the sum of the angles of a polygon.

Find the measure of each angle in an octagon.

Begin by using the formula $(n - 2) \times 180°$. An octagon has 8 sides, and so you should substitute 8 for n.

$$(8 - 2) \times 180° = 6 \times 180° = 1080°$$

Then divide the sum of the angles by the number of angles. Because an octagon has 8 angles, divide by 8.

$$1080° \div 8 = 135°$$

The answer tells you that each angle of a regular octagon measures 135°.

Check It Out

9. Find the sum of the angles of a polygon with 7 sides.
10. Find the measure of each angle in a regular hexagon.

Polyhedrons

Solid shapes can be curved, like these.

Sphere	Cylinder	Cone

Some solid shapes have flat surfaces. Each of the figures below is a **polyhedron**.

Cube	Prism	Pyramid

A polyhedron is any solid whose surface is made up of polygons. Triangles, quadrilaterals, and pentagons make up the **faces** of the common polyhedrons below.

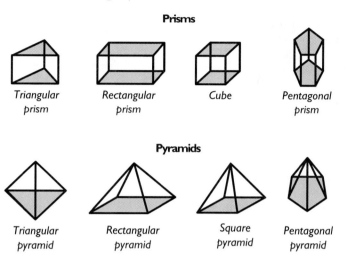

Prisms

Triangular prism   Rectangular prism   Cube   Pentagonal prism

Pyramids

Triangular pyramid   Rectangular pyramid   Square pyramid   Pentagonal pyramid

A **prism** has two bases, or "end" faces. The bases of a prism are the same size and shape and are parallel to each other. Its other faces are parallelograms. The bases of each prism in the chart are shaded. When all six faces of a **rectangular prism** are square, the figure is a **cube.**

A **pyramid** is a structure that has one base in the shape of a polygon. It has triangular faces that meet a point called the *apex.* The base of each pyramid in the chart is shaded. A triangular pyramid is a **tetrahedron.** A tetrahedron has four faces. Each face is triangular. A triangular prism, however, is *not* a tetrahedron.

 Check It Out
Identify each polyhedron.

11.

12.

7·2 EXERCISES

Use the figures to answer the questions.
1. Name the quadrilateral in two different ways.
2. Find $m\angle S$.
3. Is *PQRS* a parallelogram? Explain.

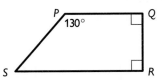

ABCD is a parallelogram.
4. What is the length of $\overline{BC}$?
5. Find $m\angle B$.
6. Is *ABCD* a rectangle? Explain.

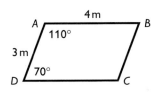

Tell whether each statement below is true or false.
7. A rectangle cannot be a rhombus.
8. Every rhombus is a parallelogram.
9. Every square is a rhombus, a rectangle, and a parallelogram.
10. Every parallelogram is a rectangle.
11. Every rectangle is a parallelogram.
12. Every trapezoid is a quadrilateral.
13. Every quadrilateral is a trapezoid.

Identify each polygon.

14.

15.

16.

17.

18.

19.

20. What is the sum of the angles in a polygon with 9 sides?
21. What is the measure of each angle in a regular pentagon?

Identify each polyhedron.

22.   23.   24.

25.   26.   27.

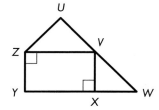

28. Name a trapezoid in the figure.
29. Name a rectangle in the figure.
30. Name a pentagon in the figure.

7·3 Symmetry and Transformations

Whenever you move a shape that is in a plane, you are performing a **transformation.**

Reflections

A **reflection** (or **flip**) is one kind of transformation. When you hear the word "reflection," you may think of a mirror. The mirror image, or reverse image, of a point or shape is called a *reflection.*

The reflection of a point is another point on the other side of a **line of symmetry.** Both the point and its reflection are the same distance from the line.

T' reflects point *T* on the other side of line *l. T'* is read "*T*-prime." *T'* is called the *image* of *T*.

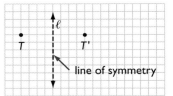

Any point, line, or polygon can be reflected. Quadrilateral *WXYZ* is reflected on the other side of line *m.* The image of *WXYZ* is *W'X'Y'Z'.*

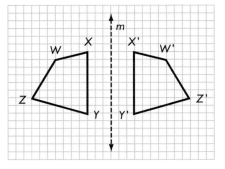

To find an image of a shape, pick several key points in the shape. For a polygon, use the vertices. For each point, measure the distance to the line of symmetry. The image of each point will be the same distance from the line of symmetry on the opposite side.

In the quadrilateral reflection on the previous page, point W is 7 units from the line of symmetry, and point W' is also 7 units from the line on the opposite side. You can measure the distance from the line for each point, and the corresponding image point will be the same distance.

 Check It Out

1. Copy the shape below on grid paper. Then find the reflection and draw and label the images.

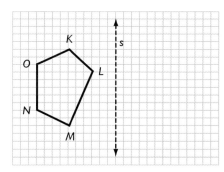

Flip It, Slide It, Turn It Messages

Can you read this secret message?
See Hot Solutions for answer.

Use flips, slides, or turns with the letters of the alphabet and with numbers to send a secret message to a friend. Make up problems for a classmate or partner to solve, using different operations, but write them in "code." See if your partner can solve them.

Reflection Symmetry

You have seen that a line of symmetry is used to show the reflection symmetry of a point, a line, or a shape. A line of symmetry can also *separate* a shape into two parts, where one part is a reflection of the other. Each of these figures is symmetrical with respect to the line of symmetry.

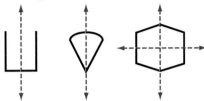

Sometimes a figure has more than one line of symmetry. Here are more shapes that have more than one line of symmetry.

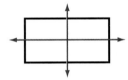

A rectangle has two lines of symmetry.

A square has four lines of symmetry.

Any line through the center of a circle is a line of symmetry. So a circle has an infinite number of lines of symmetry.

Check It Out

Tell whether each figure has reflection symmetry. If your answer is yes, tell how many lines of symmetry can be drawn through the figure.

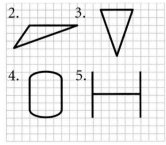

Rotations

A **rotation** (or **turn**) is a transformation that turns a line or a shape around a fixed point. This point is called the *center of rotation*. You usually measure the number of degrees of rotation of a shape in a counterclockwise direction.

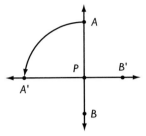

$\overleftrightarrow{AB}$ is rotated 90° around point P.

If you rotate a figure 360°, it comes back to where it started. Despite the rotation, its position is unchanged.
If you rotate $\overrightarrow{ST}$ 360° around point *M*, $\overrightarrow{ST}$ is still in the same place.

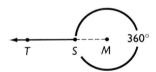

 Check It Out

6. How many degrees has $\overrightarrow{DE}$ been rotated?

7. How many degrees has $\triangle PQR$ been rotated?

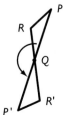

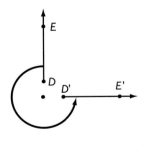

Translations

A **translation** (or **slide**) is another kind of transformation. When you slide a figure to a new position without turning it, you are performing a translation.

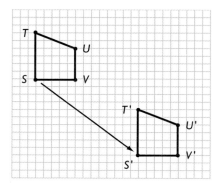

Trapezoid *STUV* moves right and down. *S'T'U'V'* is the image of *STUV* under a translation. *S'* is 13 units to the right and 10 units down from *S*. All other points on the trapezoid have moved the same way.

Check It Out

Does each pair of figures represent a translation? If you write yes, describe the translation.

8.

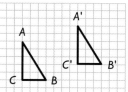

9.

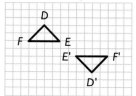

10.

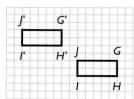

7·3 EXERCISES

Name the reflection across line *a* of each of the following.

1. point *J*   2. trapezoid *JKLM*   3. $\overline{RS}$

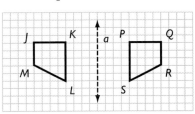

Copy the shapes. Then draw all the lines of symmetry for each.

4.   5.   6.

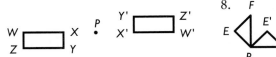

Find the number of degrees that each figure has been rotated about point *P*.

7.   8.

Find the direction and number of units that each figure has been moved in the translations.

9.   10.

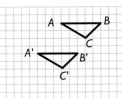   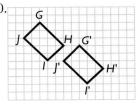

7·4 Perimeter

Perimeter of a Polygon

Dolores is planning to put a frame around a painting. To determine how much framing she needs, she must calculate the **perimeter** of, or *distance around,* the painting.

The perimeter of any polygon is the sum of the lengths of the sides of the polygon.

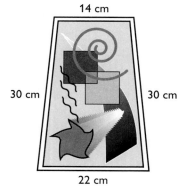

14 cm

30 cm · 30 cm

22 cm

$$P = 30 \text{ cm} + 30 \text{ cm} + 14 \text{ cm} + 22 \text{ cm} = 96 \text{ cm}$$

The perimeter of the painting is 96 cm. Dolores will need 96 cm of framing.

FINDING THE PERIMETER OF A POLYGON

To find the perimeter of any polygon, add up the lengths of all its sides.

Find the perimeter.

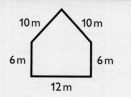

10 m   10 m

6 m   6 m

12 m

$$P = 6 \text{ m} + 10 \text{ m} + 10 \text{ m} + 6 \text{ m} + 12 \text{ m} = 44 \text{ m}$$

The perimeter of this pentagon is 44 m.

Regular Polygon Perimeters

The sides of a regular polygon are all the same length. If you know the perimeter of a regular polygon, you can find the length of each side.

To find the length of each side of a regular hexagon with a perimeter of 15 cm, let x = length of a side.

$$15 \text{ cm} = 6x \qquad\qquad 2.5 \text{ cm} = x$$

Each side is 2.5 cm long.

 Check It Out

Find the perimeter of each polygon.

1.

2.
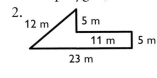

Find the length of each side.

3. a regular octagon with a perimeter of 32 ft

4. an equilateral triangle with a perimeter of 63 m

Perimeter of a Rectangle

Opposite sides of a rectangle are equal. So to find the perimeter of a rectangle, you only need to know its length and width.

FINDING THE PERIMETER OF A RECTANGLE

For a rectangle with length, l, and width, w, the perimeter, P, can be found with the formula $P = 2l + 2w$.

Find the perimeter of a rectangle with a length of 11 cm and a width of 8 cm.

$$P = 2l + 2w$$
$$= (2 \times 11) + (2 \times 8)$$
$$= 22 + 16 = 38 \text{ cm}$$

The perimeter is 38 centimeters.

A square is a rectangle whose length and width are equal. So the formula for finding the perimeter of a square, whose sides measure s, is $P = 4 \times s$ or $P = 4s$.

Check It Out

Find the perimeter.
5.   rectangle with length 10 ft and width 5 ft
6.   square with sides of 9 in

The Pentagon

Located near Washington, D.C., the Pentagon is one of the largest office buildings in the world. The United States Army, Navy, and Air Force all have their headquarters there.

The building covers an area of 29 acres and has 3,707,745 ft^2 of usable office space.

The structure consists of five concentric regular pentagons with ten spokelike corridors connecting them. The outside perimeter of the building is about 4,620 ft. What is the length of an outermost side? See Hot Solutions for answer.

Perimeter of a Right Triangle

If you know the lengths of two sides of a **right triangle,** you can find the length of the third side with the **Pythagorean Theorem.**

For a review of the *Pythagorean Theorem,* see page 379.

FINDING THE PERIMETER OF A RIGHT TRIANGLE

Use the Pythagorean Theorem to find the perimeter of the right triangle.

$a = 24$ ft
$b = 10$ ft

- Use the equation $c^2 = a^2 + b^2$ to find the length of the hypotenuse.

$$c^2 = 24^2 + 10^2$$
$$= 576 + 100$$
$$= 676$$

- The square root of c^2 is the length of the hypotenuse.

$$c = 26$$

- Add the lengths of the sides. The sum is the perimeter of the triangle.

$$24 \text{ ft} + 10 \text{ ft} + 26 \text{ ft} = 60 \text{ ft}$$

The perimeter is 60 ft.

Check It Out

Use the Pythagorean Theorem to find the perimeter of each triangle.

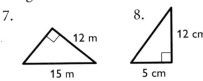

7.   12 m
15 m

8.   12 cm
5 cm

7·4 EXERCISES

Find the perimeter of each polygon.

1.

8 m
5 m · 6 m
11 m

2.

4 ft
2 ft · 2 ft
2 ft · 2 ft
4 ft

3.
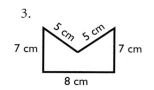
5 cm · 5 cm
7 cm · 7 cm
8 cm

4. Find the perimeter of a square with sides of 5.5 m.
5. An equilateral triangle has a perimeter of 57 cm. Find the length of each side.
6. A regular polygon has sides of 24 cm and a perimeter of 360 cm. How many sides does the polygon have?

Find the perimeter of each rectangle.
7. $l = 10$ in., $w = 7$ in.
8. $l = 23$ m, $w = 16$ m
9. $l = 6$ cm, $w = 2.5$ cm
10. $l = 32$ ft, $w = 1$ ft

Find the perimeter of each triangle.

11.
16 in.
12 in.

12.
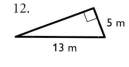
5 m
13 m

13. Each side of a regular hexagon measures 15 cm. What does the hexagon's perimeter measure?
14. The perimeter of a regular octagon is 128 in. How long is each side?
15. The perimeter of a rectangle is 40 m. Its width is 8 m. What is its length?
16. What is the length of each side of a square that has a perimeter of 66 cm?

17. Use the Pythagorean
    Theorem to find the
    perimeter of △*ABC*.

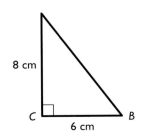

18. A town is going to put a
    fence around the park.
    How long will the
    fence be?
19. Fencing comes in rolls of
    20 m. How many rolls will
    be needed?

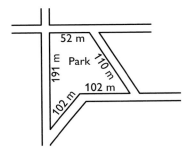

7·5 Area

What Is Area?

Area measures the size of a surface. Your desktop is a surface with area, and so is the state of Florida. Instead of measuring with units of length, such as inches, centimeters, feet, and kilometers, you measure area in square units, such as square inches (in.^{2}) and square centimeters (cm^2).

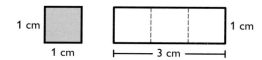

This square has an area of one square centimeter. It takes exactly three of these squares to cover this rectangle, which tells you that the area of the rectangle is three square centimeters, or 3 cm^2.

Estimating Area

When an exact answer is not needed or is hard to find, you can estimate the area of a surface.

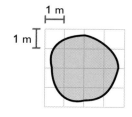

In the shaded figure to the right, four squares are completely shaded, and so you know that the area is greater than 4 m^2. The square around the shape covers 16 m^2, and obviously the shaded area is less than that. So to estimate the area of the shaded figure, you can say that it is greater than 4 m^2 but less than 16 m^2.

 Check It Out

1. Estimate the area of the shaded region. Each square represents 1 m^2.

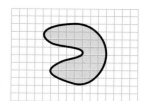

Area of a Rectangle

By counting squares you can find the area of this rectangle.

1 cm

1 cm

There are 8 squares and each is a square centimeter. So the area of this rectangle is 8 cm^2.

You can also use the formula for finding the area of a rectangle: $A = l \times w$. The length of the rectangle above is 4 cm and the width is 2 cm. Using the formula, you find that

$$A = 4 \text{ cm} \times 2 \text{ cm}$$
$$= 8 \text{ cm}^2$$

FINDING THE AREA OF A RECTANGLE

Find the area of this rectangle.

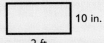

10 in.

2 ft

- The length and the width must be in the same units.

  2 ft = 24 in. So $l = 24$ in. and $w = 10$ in.

- Use the formula for the area of a rectangle.

  $$A = l \times w$$
  $$= 24 \text{ in.} \times 10 \text{ in.}$$
  $$= 240 \text{ in.}^2$$

The area of the rectangle is 240 in.^{2}.

If the rectangle is a square, the length and the width are the same. So for a square whose sides measure s units, you can use the formula $A = s \times s$, or $A = s^2$.

7·5 AREA

Check It Out

2. Find the area of a rectangle with a length of 3 ft and a width of 9 in.
3. Find the area of a square whose sides measure 13 m.

Area of a Parallelogram

To find the area of a parallelogram, you multiply the base by the height.

Area = base × height
$A = b \times h$
or $A = bh$

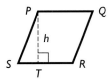

The height of a parallelogram is always **perpendicular** to the base. So in parallelogram $PQRS$, the height, h, is equal to $\overline{PT}$, not $\overline{PS}$. The base, b, is equal to $\overline{SR}$.

FINDING THE AREA OF A PARALLELOGRAM

Find the area of a parallelogram with a base of 9 in. and a height of 5 in.

$A = b \times h$
$ = 9 \text{ in.} \times 5 \text{ in.}$
$ = 45 \text{ in.}^2$

The area of the parallelogram is 45 in.^{2} or 45 sq in.

Check It Out

4. Find the area of a parallelogram with a base of 6 cm and a height of 8 cm.
5. Find the height of a parallelogram that has a base of 12 ft and an area of 132 ft^2.

Area of a Triangle

If you were to cut a parallelogram along a diagonal, you would have two triangles with equal bases, *b*, and the same height, *h*.

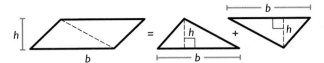

A triangle has half the area of a parallelogram with the same base and height. The area of a triangle equals $\frac{1}{2}$ the base times the height, and so the formula is $A = \frac{1}{2} \times b \times h$, or $A = \frac{1}{2}bh$.

$A = \frac{1}{2} \times b \times h$
$A = \frac{1}{2} \times 7.5 \times 8.2$
$\quad = 0.5 \times 7.5 \times 8.2$
$\quad = 30.75 \text{ m}^2$

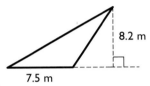

The area of the triangle is 30.75 m².

FINDING THE AREA OF A TRIANGLE

Find the area of $\triangle PQR$. Note that in a right triangle, the two **legs** serve as a height and a base.

$A = \frac{1}{2}bh$
$\quad = \frac{1}{2} \times 4 \times 6$
$\quad = 0.5 \times 4 \times 6$
$\quad = 12 \text{ ft}^2$

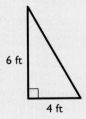

The area of the triangle is 12 ft².
For a review of *right triangles,* see page 378.

Check It Out

6. Find the area of a triangle with a base of 15 cm and a height of 10 cm.
7. Find the area of a right triangle whose sides measure 10 in., 24 in., and 26 in.

Area of a Trapezoid

A trapezoid has two bases, which are labeled b_1 and b_2. You read b_1 as "b sub-one." The area of a trapezoid is equal to the area of two triangles.

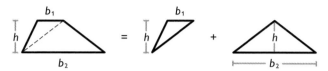

You know that the formula for the area of a triangle is $A = \frac{1}{2}bh$, and so it makes sense that the formula for finding the area of a trapezoid would be $A = \frac{1}{2}b_1h, + \frac{1}{2}b_2h$ or, in simplified form, $A = \frac{1}{2}h(b_1 + b_2)$.

FINDING THE AREA OF A TRAPEZOID

Find the area of trapezoid $ABCD$.

$$A = \tfrac{1}{2}h(b_1 + b_2)$$
$$= \tfrac{1}{2} \times 6 \, (8 + 12)$$
$$= 3 \times 20$$
$$= 60 \text{ m}^2$$

The area of the trapezoid is 60 m².

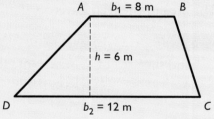

Because $\frac{1}{2}h(b_1 + b_2)$ is equal to $h \times \frac{b_1 + b_2}{2}$, you can remember the formula this way:

   $A =$ height times the average of the bases

For a review of how to find an *average* or *mean*, see page 210.

Check It Out

8. The height of a trapezoid is 5 m. The bases are 3 m and 7 m. What is the area?

9. The height of a trapezoid is 2 cm. The bases are 8 cm and 9 cm. What is the area?

7·5 EXERCISES

1. Estimate the area of the shaded part of the figure below.

2. If each square unit in the figure is 2 cm², estimate the area in centimeters².

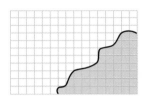

Find the area of each rectangle, with length, *l,* and width, *w.*
  3. *l* = 14 in., *w* = 7 in.   4. *l* = 19 cm, *w* = 1 m

Find the area of each parallelogram.

5.   6.

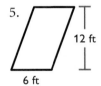

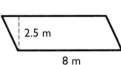

Find the area of each triangle, given base, *b,* and height, *h.*
  7. *b* = 16 cm, *h* = 10 cm   8. *b* = 4 ft, *h* = 3.5 ft

9. A trapezoid has a height of 2 ft. Its bases measure 6 in. and 1 ft. What is its area?
10. Find the area of the figure.

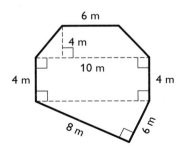

7·6 Surface Area

The **surface area** of a solid is the total area of its exterior surfaces. You can think about surface area in terms of the parts of a solid shape that you would paint. Like area, surface area is expressed in square units. To see why, "unfold" the rectangular prism.

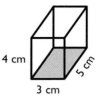

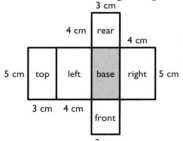

Mathematicians call this unfolded prism a *net*. A net is a pattern that can be folded to make a three-dimensional figure.

Surface Area of a Rectangular Prism

A rectangular prism has six rectangular faces. To find the surface area of a rectangular prism, find the sum of the areas of the six faces, or rectangles. Remember, opposite faces are equal. For a review of *polyhedrons* and *prisms*, see page 340.

SURFACE AREA OF A RECTANGULAR PRISM

Use the net to find the area of the rectangular prism above.

- Use the formula $A = lw$ to find the area of each face.
- Then add the six areas.
- Express the answer in square units.

$$
\begin{aligned}
\text{Area} &= \text{top} + \text{base} &+& \quad \text{left} + \text{right} &+& \quad \text{front} + \text{rear} \\
&= 2 \times (3 \times 5) &+& \quad 2 \times (4 \times 5) &+& \quad 2 \times (3 \times 4) \\
&= 2 \times 15 &+& \quad 2 \times 20 &+& \quad 2 \times 12 \\
&= 30 &+& \quad 40 &+& \quad 24 \\
&= 94 \text{ cm}^2
\end{aligned}
$$

The surface area of the rectangular prism is 94 cm^2.

Check It Out
Find the surface area of each shape.

1.

10 in.

5 in.   5 in.

2.

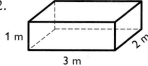

1 m

2 m

3 m

Surface Area of Other Solids

The unfolding technique can be used to find the surface area of any polyhedron. Look at the **triangular prism** and its net.

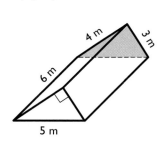

4 m

3 m

6 m

5 m

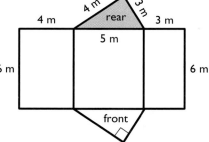

4 m

4 m   rear   3 m

3 m

5 m

6 m   6 m

front

To find the surface area of this solid, use the area formulas for a rectangle ($A = lw$) and a triangle ($A = \frac{1}{2}bh$). These will find the areas of the five faces. Then find the sum of the areas.

Below are two pyramids and their nets. For these polyhedrons, you would again use the area formulas for a rectangle ($A = lw$) and a triangle ($A = \frac{1}{2}bh$). These will find the areas of the faces and then you can find the sum of the areas.

Rectangular pyramid

Tetrahedron (triangular pyramid)

The surface area of a **cylinder** is the sum of the areas of two **circles** and a rectangle.

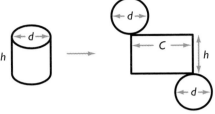

The two bases of a cylinder are equal in area. The height of the rectangle is the height of the cylinder. Its length is the **circumference** of the cylinder.

To find the surface area of a cylinder, you would:
- Use the formula for the area of a circle to find the area of each base.

$$A = \pi r^2$$

- Find the area of the rectangle with the formula $h \times (2\pi r)$.

For a review of *circles,* see page 372.

Check It Out

3. Unfold the triangular prism and find its surface area.
4. Which unfolded figure represents the pyramid?

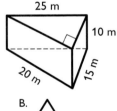

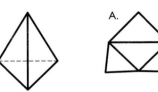

A.

B.

5. Find the surface area of the cylinder. Use $\pi = 3.14$.

7·6 EXERCISES

1. Find the surface area of a rectangular prism with sides of 2 m, 3 m, and 6 m.
2. Find the surface area of a rectangular prism with sides of 1 in., 12 in., and 12 in.
3. Find the surface area of a cube with sides of 1 in.
4. A cube has a surface area of 150 m². Find the length of each side.

Find the surface area of each triangular prism.

5.

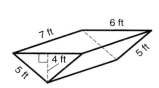

6.

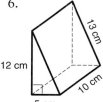

7. Which net shows a regular tetrahedron?

A.

B.

Find the surface area of each cylinder. Use π = 3.14.

8.

9.

10. Find the surface area of the square pyramid. All the triangular faces are congruent.

7·7 Volume

What Is Volume?

Volume is the space inside a figure. One way to measure volume is to count the number of cubic units that would fill the space inside a figure.

The volume of this small cube is 1 cubic inch.

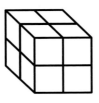

The number of smaller cubes that it takes to fill the space inside the larger cube is 8, and so the volume of the larger cube is 8 cubic inches.

You measure the volume of shapes in *cubic* units. For example, 1 cubic inch is written as 1 in.^{3}, and 1 cubic meter is written as 1 m^3.

For a review of *cubes*, see page 340.

Check It Out

What is the volume of each shape?

1.   1 cube = 1 m^3

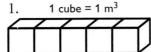

2. 1 cube = 1 cm^3

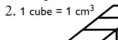

Volume of a Prism

The volume of a prism can be found by multiplying the *area* (pp. 356–360) of the base, *B*, and the height, *h*.

Volume = *Bh*

See *formulas*, pages 60–61.

FINDING THE VOLUME OF A PRISM

Find the volume of the rectangular prism. The base is 5 m long and 4 m wide; the height is 10 m.

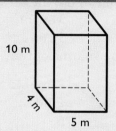

base *B* = 5 m × 4 m
   = 20 m²
V = 20 m² × 10 m
   = 200 m³

• Find the area of the base.

• Multiply the base and the height.

The volume of the prism is 200 m³.

 Check It Out

Find the volume of each shape.

3.

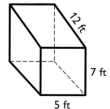

4.

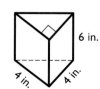

7·7 VOLUME

Volume of a Cylinder

You can find the volume of a cylinder the same way you found the volume of a prism, using the formula $V = Bh$. *Remember:* The base of a cylinder is a circle.

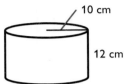

10 cm

12 cm

The base has a radius of 10 cm, so the area (πr^2) is $\pi \times$ 100 cm^2, about 314 cm^2. Because you also know the height, you can use the formula $V = Bh$.

$V = 314 \text{ cm}^2 \times 12 \text{ cm}$
$= 3{,}768 \text{ cm}^3$

The volume of the cylinder is 3,768 cm^3.

 Check It Out

Find the volume of each cylinder. Use $\pi = 3.14$.

5.

20 ft

50 ft

6.

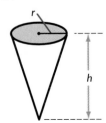

10 cm

20 cm

Volume of a Pyramid and a Cone

The formula for the volume of a pyramid or a cone is $V = \frac{1}{3}Bh$.

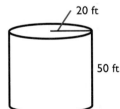

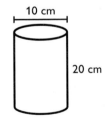

r

h

FINDING THE VOLUME OF A PYRAMID

You can find the volume of a pyramid using the formula $V = \frac{1}{3}Bh$.

B is the area of the base of the pyramid, and h is the height of the pyramid.

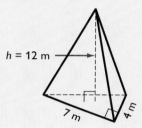

$h = 12$ m

7 m

4 m

$B = \frac{1}{2} \times 4 \text{ m} \times 7 \text{ m}$ • Find the area of the base.

$= 14 \text{ m}^2$

$V = \frac{1}{3}Bh$ • Multiply the area of the

$= \frac{1}{3} \times 14 \text{ m}^2 \times 12 \text{ m}$ base by the height and

$= 56 \text{ m}^3$ by $\frac{1}{3}$.

The volume of the pyramid is 56 m^3.

To find the volume of a cone, you follow the same procedure as above. You may use your calculator to help find the area of the base of the cone. For example, a cone has a base with a radius of 3 cm and a height of 10 cm. What is the volume of the cone to the nearest tenth?

Square the radius and multiply by π to find the area of the base. Then multiply by the height and divide by 3 to find the volume. The volume of the cone is 94.2 cm^3.

Press $\boxed{\pi}$ $\boxed{\times}$ 9 $\boxed{=}$ $\boxed{28.27433}$ $\boxed{\times}$ 10 $\boxed{\div}$ 3 $\boxed{=}$ $\boxed{94.24778}$

For other volume *formulas,* see pages 60–61.

Check It Out

Find the volume of the shapes below, rounded to the nearest tenth.

7.

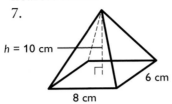

$h = 10$ cm

6 cm

8 cm

8.

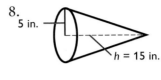

5 in.

$h = 15$ in.

Good Night, T. Rex

Why did the dinosaurs disappear? New evidence from the ocean floor points to a giant asteroid that collided with Earth some 65 million years ago.

The asteroid, 6 to 12 mi in diameter, hit the earth somewhere in the Gulf of Mexico. It was traveling at a speed of thousands of miles per hour.

The collision sent billions of tons of debris into the atmosphere. The debris rained down on the planet, obscuring the sun. Global temperatures plummeted. The fossil record shows that most of the species that were alive before the collision disappeared.

Assume the crater left by the asteroid had the shape of a hemisphere with a diameter of 165 mi. How many cubic miles of debris would have been flung from the crater into the air? For formula for volume of sphere, see p. 60. See Hot Solutions for answer.

7·7 EXERCISES

1. A rectangular prism has sides of 4 in., 9 in., and 12 in. Find the volume of the prism.
2. The volume of a rectangular prism is 140 ft^3. The length of the base is 7 ft and the width of the base is 4 ft. What is the height of the prism?
3. Find the volume of a cube with sides of 10 cm.
4. A cube has a volume of 125 m^3. What is the length of each side of the cube?

Find the volume of each solid. Use $\pi = 3.14$.

5.
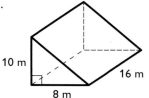
10 m · 8 m · 16 m

6.
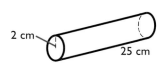
2 cm · 25 cm

7.
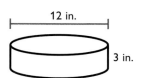
12 in. · 3 in.

8.
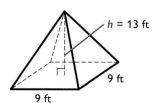
h = 13 ft · 9 ft · 9 ft

9.
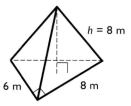
h = 8 m · 6 m · 8 m

10.
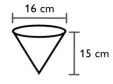
16 cm · 15 cm

7·8 Circles

Parts of a Circle

Of the many shapes you may encounter in geometry, circles are among the most unique. They differ from other geometric shapes in several ways. For instance, although all circles are the same shape, polygons vary in shape. Circles do not have any sides, while polygons are named and classified by the number of sides they have. The *only* thing that makes one circle different from another is size.

A circle is a set of points equidistant from a given point. That point is the center of the circle. A circle is named by its center point.

Circle S

A **radius** is a **segment** that has one endpoint at the center and the other endpoint on the circle. In circle S, $\overline{SL}$ is a *radius*, and so is $\overline{SK}$.

A **diameter** is a line segment that passes through the center of the circle and has both endpoints on the circle. $\overline{KL}$ is a diameter of circle S. Notice that the length of the diameter $\overline{KL}$ is equal to the sum of $\overline{SK}$ and $\overline{SL}$. So the diameter is twice the length of the radius. If d equals the diameter and r equals the radius, d is twice the radius, r. So the diameter of circle S is 2(7) or 14 cm.

Check It Out

1. Find the radius of a circle with diameter 26 m.
2. Find the radius of a circle with diameter 1 cm.
3. Find the diameter of a circle in which $r = 16$ in.
4. Find the diameter of a circle in which $r = 2.5$ ft.
5. The diameter of circle P measures twice the diameter of circle Q. The radius of circle Q measures 6 m. What is the length of the radius of circle P?

Circumference

The circumference of a circle is the distance around the circle. The ratio (p. 292) of every circle's circumference to its diameter is always the same. That ratio is a number close to 3.14. In other words, in every circle, the circumference is about 3.14 times the diameter. The symbol π, which is read as **pi,** is used to represent the ratio $\frac{c}{d}$.

$$\frac{c}{d} = 3.141592 \dots$$

Circumference = pi $\times$ diameter, or $C = \pi d$

Look at the illustration below. The circumference of the circle is about the same length as three diameters. This is true for any circle.

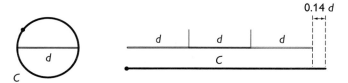

Because $d = 2r$, Circumference = two $\times$ pi $\times$ radius, or $C = 2\pi r$.

If you are using a calculator that has a π key, hit it, and you will get an approximation for π to several decimal places: $\pi = 3.141592 \dots$. For practical purposes, however, when you are finding the circumference of a circle, round π to 3.14, or simply leave the answer in terms of π.

7·8 CIRCLES

FINDING THE CIRCUMFERENCE OF A CIRCLE

Find the circumference, to the nearest tenth, of a circle with radius 6 m.

$d = 6$ m $\times 2 = 12$ m
- To get the diameter, remember to multiply the radius by 2.

$C = \pi \times 12$ m
- Use the formula $C = \pi d$.

The exact circumference is 12π m.
- You can leave the answer in terms of π.

$C = 3.14 \times 12$ m
- Or, use $\pi = 3.14$.

$\quad = 37.68$ m ≈ 37.7 m
- Round your answer, if necessary.

To the nearest tenth, the circumference is 37.7 m.

You can find the diameter of a circle if you know its circumference. $d = \frac{C}{\pi}$

 Check It Out

Give your answers in terms of π.

6. What is the circumference of a circle with radius 14 ft?

7. What is the circumference of a circle with diameter 21 cm?

Use $\pi = 3.14$. Round your answers to the nearest tenth.

8. Find the circumference of a circle with diameter 14.6 m.

9. Find the circumference of a circle with radius 18 cm.

10. Find the diameter of a circle that has a circumference of 20.41 in.

Central Angles

A central angle is an angle whose vertex is at the center of a circle. The sum of the central angles in any circle is 360°.

For a review of *angles,* see page 327.

The part of a circle where a central angle intercepts the circle is called an **arc.** The measure of the arc, in degrees, is equal to the measure of the central angle.

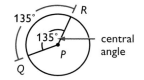

$m\angle QPR = 135°$ and $\overset{\frown}{QR} = 135°$

 Check It Out

11. Name a central angle of circle *X.*
12. What is the measure of $\overset{\frown}{WY}$?
13. What is the measure of $\angle MPN$?

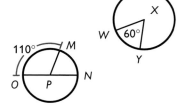

Now, That's a Pizza!

Put together 10,000 lb of flour, 664 gal of water, 316 gal of tomato sauce, 1,320 lb of cheese, and 1,200 lb of pepperoni. What have you got? An 18,664-lb pizza, baked by L. Amato in 1978 to raise money for charity.

Together with L. Piancone, Amato organized the baking of another giant pizza in 1991. This pie, with a diameter of about $56\frac{1}{2}$ ft, still holds the record for the largest pizza made in the United States. What was its area? See Hot Solutions for answer.

7·8 CIRCLES

Area of a Circle

To find the area of a circle, you use the formula Area = pi × radius^2, or $A = \pi r^2$. As with the area of polygons, the area of a circle is expressed in square units.

For a review of *area* and *square units,* see page 356.

FINDING THE AREA OF A CIRCLE

Find the area of circle *T* to the nearest whole number.

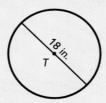

$r = 18$ in. $\div 2 = 9$ in. • To find the radius, divide the diameter by 2.

$A = \pi \times 9$ in.$^2 = 81\pi$ in.^{2} • Use the formula $A = \pi r^2$.

≈ 254.34 in.^{2} • Use $\pi = 3.14$.

≈ 254 in.^{2} • Round to the nearest whole number.

The area of circle *T* is about 254 in.^{2}

Check It Out

14. Find the area of a circle with radius 13 cm. Give your answer in terms of π.

15. A circle has a diameter of 21 ft. Find the area of a circle to the nearest whole number. Use $\pi = 3.14$.

7·8 EXERCISES

1. What is the radius of a circle with diameter 42.8 m?
2. The diameter of a circle is 10 cm greater than the radius. How long is the radius?

Find the circumference to the nearest tenth of each circle with given radius or diameter. Use $\pi = 3.14$.

3. $d = 15$ in.   4. $d = 7$ m   5. $r = 5.5$ cm
6. The circumference of a circle measures 63.4 ft. Find the circle's diameter to the nearest tenth.
7. Find the radius to the nearest whole number of a circle that has a circumference of 1,298 m.

Use circle S to answer items 8–11.

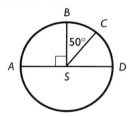

8. Find the measure of arc AC.
9. Find the measure of $\angle CSD$.
10. Find the measure of arc AD.
11. Name two arcs that have a measure of 90°.

Find the area in terms of π of each circle with given radius or diameter.

12. $r = 11$ ft   13. $d = 60$ cm
14. $r = 1.5$ in.

Find the area to the nearest tenth of each circle with given radius or diameter. Use $\pi = 3.14$.

15. $d = 16$ m   16. $r = 9$ ft
17. $r = 12.8$ cm

18. A circle has a circumference of 25 in. Find the area of the circle to the nearest whole number.
19. If you double the diameter of a circle, you increase its circumference by
    A. 2 times   B. π times   C. 2^2 times
20. If you double the radius of a circle, you increase its area by
    A. 2 times   B. π times   C. 2^2 times

7·9 Pythagorean Theorem

Right Triangles

The illustration below left shows a right triangle on a geoboard. The triangle has an area of $\frac{1}{2}$ square unit and each leg is one unit long.

Now look at the squares on each of the three sides of the triangle. Call the squares *A*, *B*, and *C*.

Area $A = 1 \times 1 = 1$ square unit
Area $B = 1 \times 1 = 1$ square unit

When you look at the pegs of the geoboard, the area of *C* is not easy to determine, but it is clear that it is equal to four of the original triangles.

Area $C = 4 \times \frac{1}{2}$
$\qquad C = 2$

The area of *C* is 2 square units.

Note the relationship among the three areas:
Area A + Area B = Area C
This relationship holds true for all right triangles.

 Check It Out

1. What is the area of each of the squares?
2. What is the relationship among the squares?

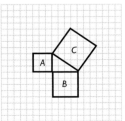

The Pythagorean Theorem

When you look at the areas of the squares on each of the sides of the triangles shown on the geoboards, you see the relationship of the area of the square of the **hypotenuse,** the side opposite the right angle, to the areas of the squares of the legs. That relationship, or pattern, is based on the lengths of all three legs. A Greek mathematician named Pythagoras noticed the relationship about 2,500 years ago and drew a conclusion. That conclusion, known as the Pythagorean Theorem, can be stated as follows: In a right triangle, the square of the length of the hypotenuse is equal to the sum of the squares of the lengths of the legs.

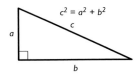

You can use the Pythagorean Theorem to find the third side of a right triangle if you know two sides.

USING THE PYTHAGOREAN THEOREM TO FIND THE HYPOTENUSE

Use the Pythagorean Theorem to find the length of the hypotenuse, c, of $\triangle WXY$.

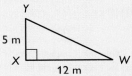

$c^2 = a^2 + b^2$ • Substitute known lengths for a and b.

$c^2 = 5^2 + 12^2$ • Square the two known lengths.

$c^2 = 25 + 144$ • Find the sum of the squares of the two legs.

$c^2 = 169$

$c = 13$ • Take the square root of the sum.

The hypotenuse measures 13 m.

USING THE PYTHAGOREAN THEOREM TO FIND A SIDE LENGTH

Use the Pythagorean Theorem, $c^2 = a^2 + b^2$, to find the length of the unknown leg, b, of a right triangle with a hypotenuse of 10 cm and one leg measuring 4 cm. Round to nearest tenth of a centimeter.

$10^2 = 4^2 + b^2$ • Use $c^2 = a^2 + b^2$.

$100 = 16 + b^2$ • Square the known lengths.

$100 - 16 = (16 - 16) + b^2$ • Subtract to isolate unknown.

$84 = b^2$

$9.16515 \ldots = b$ • Use your calculator to find the square root.

9.2 cm is the length of the unknown side.

Check It Out

3. Find the length to the nearest whole number of the hypotenuse of a right triangle with legs measuring 5 m and 6 m.

4. Find the length of $\overline{KL}$ to the nearest whole number.

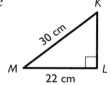

Pythagorean Triples

The numbers 3, 4, and 5 form a **Pythagorean triple** because $3^2 + 4^2 = 5^2$. Pythagorean triples are formed by whole numbers, so that $a^2 + b^2 = c^2$. There are many Pythagorean triples. Here are three:

5, 12, 13   8, 15, 17   7, 24, 25

If you multiply each number of a Pythagorean triple by the same number, you form another Pythagorean triple. 6, 8, 10 is a triple because it is 2(3), 2(4), 2(5).

7·9 EXERCISES

1. What is the relationship among the lengths *x*, *y*, and *z*?

Find the missing length in each right triangle. Round to the nearest tenth.

2.

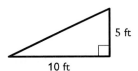

5 ft

10 ft

3.

12 cm

14 cm

4.

7 in.

7 in.

5.

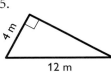

4 m

12 m

Are the following sets of numbers Pythagorean triples? Write Yes or No.

6. 9, 12, 15   7. 10, 24, 26   8. 8, 16, 20

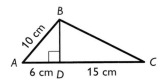

B

10 cm

A   6 cm D   15 cm   C

9. Find the length of $\overline{BD}$.
10. Find the length of $\overline{BC}$.

7·10 Tangent Ratio

Sides and Angles in a Right Triangle

In every right triangle there is one right angle and two **acute angles.** The hypotenuse, which is the longest side, is opposite the right angle. The other two sides are called the *legs of a triangle.*

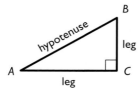

$m\angle A$ and $m\angle B < 90°$

Sometimes the legs are called the *opposite* and *adjacent* sides to describe where they are in relation to one of the acute angles of the right triangle.

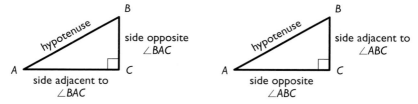

Tangent of an Angle

For any acute angle of a right triangle, the ratio of the length of the opposite side and the length of the adjacent side is called the **tangent** of the angle.

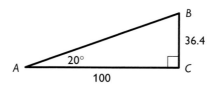

Tangent of $\angle BAC = \dfrac{BC}{AC} = \dfrac{36.4}{100} = 0.364$

Tangent Table

Because the tangent ratio always remains the same for an angle measure, you will find a table of tangents helpful when you are solving problems with right triangles.

To find the tangent of an angle, find the measure of the angle in the Angle column and then read the tangent ratio in the Tangent column.

Angle	Tangent	Angle	Tangent	Angle	Tangent
1°	0.0175	31°	0.6009	61°	1.8040
2°	0.0349	32°	0.6249	62°	1.8807
3°	0.0524	33°	0.6494	63°	1.9626
4°	0.0699	34°	0.6754	64°	2.0503
5°	0.0875	35°	0.7002	65°	2.1445
6°	0.1051	36°	0.7265	66°	2.2460
7°	0.1228	37°	0.7536	67°	2.3559
8°	0.1405	38°	0.7813	68°	2.4751
9°	0.1584	39°	0.8098	69°	2.6051
10°	0.1763	40°	0.8391	70°	2.7475
11°	0.1944	41°	0.8693	71°	2.9042
12°	0.2126	42°	0.9004	72°	3.0777
13°	0.2309	43°	0.9325	73°	3.2709
14°	0.2493	44°	0.9657	74°	3.4874
15°	0.2679	45°	1.0000	75°	3.7321
16°	0.2867	46°	1.0355	76°	4.0108
17°	0.3057	47°	1.0724	77°	4.3315
18°	0.3249	48°	1.1106	78°	4.7046
19°	0.3443	49°	1.1504	79°	5.1446
20°	0.3640	50°	1.1918	80°	5.6713
21°	0.3839	51°	1.2349	81°	6.3138
22°	0.4040	52°	1.2799	82°	7.1154
23°	0.4245	53°	1.3270	83°	8.1443
24°	0.4452	54°	1.3764	84°	9.5144
25°	0.4663	55°	1.4281	85°	11.4301
26°	0.4877	56°	1.4826	86°	14.3007
27°	0.5095	57°	1.5399	87°	19.0811
28°	0.5317	58°	1.6003	88°	28.6363
29°	0.5543	59°	1.6643	89°	57.2900
30°	0.5774	60°	1.7321		

7·10 TANGENT RATIO

Using the Tangent Ratio

You can use the tangent ratio to find the measure of an unknown side or angle in a right triangle.

Find the length of $\overline{PR}$.

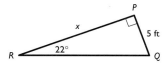

Opposite $= \overline{PQ}$ • Identify the side opposite and
Adjacent $= \overline{PR}$ adjacent to the known angle.

Tan $22° = \frac{5}{x}$ • Write the tangent ratio as a fraction

$0.4040 = \frac{5}{x}$ • Use tangent tables to find tan$R$.

$x = \frac{5}{0.4040}$ • Cross multiply and solve for x.

$x = 12.38$ • Simplify.

The length of $\overline{PR}$ is 12.38 ft.

Find $m\angle STU$.

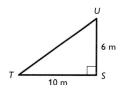

Opposite $= \overline{US}$ • Identify the side opposite and
Adjacent $= \overline{TS}$ adjacent to the angle.

Tan $T = \frac{6}{10}$ • Write the tangent ratio as a
             fraction.

$= 0.6$ • Find the decimal equivalent of
            the fraction.

Tan $31° \approx 0.6$ • Use tangent tables to find the
             angle with the tangent you have
             found.

$m\angle STU = 31°$

 Check It Out

1. Find the length of $\overline{DE}$.
2. Find $m\angle WXY$.

7·10 EXERCISES

Find the value of each tangent to the nearest hundredth.

1. tan *G*
2. tan *H*
3. tan *L*
4. tan *N*

In each right triangle, find the length of *x* to the nearest tenth.

5.

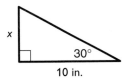

6.

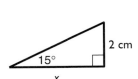

7.

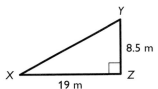

8.

In each right triangle, find the measure of angle *X* to the nearest degree.

9.

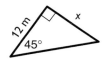

10.

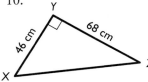

What have you learned?

You can use the problems and list of words below to see what you have learned in this chapter. You can find out more about a particular problem or word by referring to the boldfaced topic number (for example, **7•2**).

Problem Set

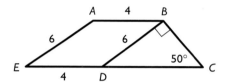

1. Name three obtuse angles in the figure. **7•1**
2. Find $m\angle BDC$. **7•1**
3. What kind of quadrilateral is *ABDE*? **7•2**
4. What do you call a rhombus that is a rectangle? **7•2**
5. What is the sum of the angles in an octagon? **7•2**
6. Write the letters of your first and last names. Do any letters have more than one line of symmetry? If so, which ones? **7•3**
7. What is the length of each side of a regular hexagon that has a perimeter of 132 cm? **7•4**
8. Find the perimeter of a rectangle with length 17 m and width 6 m. **7•4**
9. Find the area of a triangle with base 9 ft and height 14 ft. **7•5**
10. The bases of a trapezoid measure 9 in. and 17 in. Its height is 10 in. Find the area of the trapezoid. **7•5**
11. Find the surface area of a rectangular prism that has sides measuring 5 cm, 6 cm, and 7 cm. **7•6**
12. A cylinder with a radius of 3 m is 5 m high. Find the surface area of the cylinder to the nearest tenth. Use $\pi = 3.14$. **7•6**
13. A cube has a volume of 64 in³. Find the length of each side of the cube. **7•7**
14. The triangular ends of a triangular prism each have base 10 cm and height 7 cm. The prism is 5 cm long. What is its volume? **7•7**

15. The height of a rectangular pyramid is 14 cm. Its base measures 7 cm by 11 cm. What is the volume of the pyramid? **7•7**

16. Find the measure of arc *RP*. **7•8**

17. What is the circumference of circle *S* in terms of π? **7•8**

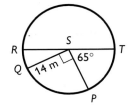

18. Find the length of *KM*. **7•9**

19. Find the length of *KN*. **7•9**

20. What is the value of tangent ∠*MKL*? **7•10**

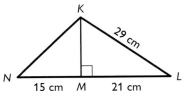

WRITE DEFINITIONS FOR THE
FOLLOWING WORDS.

hot **words**

line **7•1**
line of symmetry **7•3**
opposite angle **7•2**
parallel **7•2**
parallelogram **7•2**
pentagon **7•2**
perimeter **7•4**
perpendicular **7•5**
pi **7•8**
point **7•1**
polygon **7•1**
polyhedron **7•2**
prism **7•2**
pyramid **7•2**
Pythagorean Theorem **7•4**
Pythagorean triple **7•9**
quadrilateral **7•2**
radius **7•8**

acute angle **7•10**
angle **7•1**
arc **7•8**
circle **7•6**
circumference **7•6**
congruent **7•1**
cube **7•2**
cylinder **7•6**
degree **7•1**
diagonal **7•2**
diameter **7•8**
face **7•2**
hexagon **7•2**
hypotenuse **7•9**
isosceles triangle **7•1**
legs of a triangle **7•5**

ray **7•1**
rectangular prism **7•2**
reflection **7•3**
regular shape **7•2**
rhombus **7•2**
right angle **7•1**
right triangle **7•4**
rotation **7•3**
segment **7•8**
surface area **7•6**
tangent **7•10**
tetrahedron **7•2**
transformation **7•3**
translation **7•3**
trapezoid **7•2**
triangular prism **7•6**
vertex **7•1**
volume **7•7**

Measurement

What do you already know?

You can use the problems and list of words below to see what you already know about this chapter. The answers to the problems are in Hot Solutions at the back of the book, and the definitions of the words are in Hot Words at the front of the book. You can find out more about a particular problem or word by referring to the boldfaced topic number (for example, **8•2**).

Problem Set

1. Name one metric unit and one customary unit of weight.   **8•1**

2. Name one metric unit and one customary unit of distance.   **8•1**

3. Which system of measurement uses acres and gallons?   **8•1**

Complete the conversions.   **8•2**
4. 4.5 km = ? m
5. 8.4 m = ? mm
6. 4 yd = ? in.
7. 1,320 ft = ? mi

Use the rectangle to answer items 8–11.

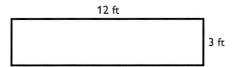

12 ft

3 ft

8. What is the figure's perimeter in inches?   **8•2**
9. What is the figure's perimeter in yards?   **8•2**
10. What is the figure's area in square inches?   **8•3**
11. What is the figure's area in square yards?   **8•3**

Complete the conversions.   **8•3**
12. $23.8 \text{ cm}^2 = ? \text{ mm}^2$
13. $7 \text{ yd}^2 = ? \text{ ft}^2$
14. $1 \text{ yd}^2 = ? \text{ in.}^2$
15. $1.9 \text{ m}^3 = ? \text{ cm}^3$
16. $216 \text{ ft}^3 = ? \text{ yd}^3$

Use the rectangular prism to answer items 17–19. **8•3**

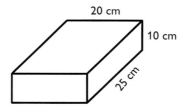

17. What is the prism's volume in cubic millimeters?
18. What is the prism's volume in cubic meters?
19. If 1 in. = 2.54 cm, what is the approximate volume of the prism in cubic inches?

20. A recipe calls for 10 oz of rice for 4 people. How many pounds of rice do you need to make the recipe for 24 people? **8•4**
21. How many 200-g packets of dried beans would you have to buy if you needed 2.4 kg of beans? **8•4**
22. How many minutes are there in 3 weeks? **8•5**
23. How many seconds are there in a year? **8•5**

A photograph that was 9 in. long by 12 in. high was reduced so that the new photo was 8 in. high. **8•6**
24. What is the new length of the photo?
25. What is the ratio of the area of the new photo to the area of the original photo?

WHAT DO YOU KNOW?

CHAPTER 8		
hot **words**	factor **8•1**	rounding **8•1**
	fraction **8•1**	scale factor **8•6**
	length **8•2**	side **8•1**
accuracy **8•1**	metric system **8•1**	similar figures **8•6**
area **8•1**	perimeter **8•1**	
customary system **8•1**	power **8•1**	square **8•1**
distance **8•2**	ratio **8•6**	time **8•5**
	rectangle **8•1**	volume **8•3**

8·1 Systems of Measurement

If you have ever watched the Olympic Games, you may have noticed that the distances are measured in meters or kilometers, and weights are measured in kilograms. That is because the most common system of measurement in the world is the **metric system.** In the United States, we use the **customary system** of measurement. It will be useful for you to be able to make conversions from one unit of measurement to another within each system, as well as convert units between the two systems.

The Metric and Customary Systems

The metric system of measuring is based on **powers** of ten, such as 10, 100, and 1,000. Converting within the metric system is simple because it is easy to multiply and divide by powers of ten.

Prefixes in the metric system have consistent meanings.

Prefix	Meaning	Example
milli-	one thousandth	1 *milli*liter is 0.001 liter.
centi-	one hundredth	1 *centi*meter is 0.01 meter.
kilo-	one thousand	1 *kilo*gram is 1,000 grams.

BASIC MEASURES

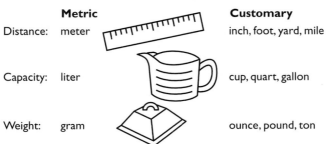

	Metric	**Customary**
Distance:	meter	inch, foot, yard, mile
Capacity:	liter	cup, quart, gallon
Weight:	gram	ounce, pound, ton

The customary system of measurement is not based on powers of ten. It is based on numbers like 12 and 16, which have many **factors.** This makes it easy to find, say, $\frac{2}{3}$ ft or $\frac{3}{4}$ lb. While the metric system uses decimals, you will frequently encounter **fractions** in the customary system.

Unfortunately there are no convenient prefixes as in the metric system, so you will have to memorize the basics: 16 oz = 1 lb; 36 in. = 1 yd; 4 qt = 1 gal; and so on.

Check It Out

1. Which system is based on multiples of 10?
2. Which system uses fractions?

8·1 SYSTEMS OF MEASUREMENT

From Boos to Cheers

It took 200 skyjacks two years and 2.5 million rivets to put together the Eiffel Tower. When it was completed in 1899, the art critics of Paris considered it a blight on the landscape. Today, it is one of the most familiar and beloved monuments in the world.

The tower's height, not counting its TV antennas, is 300 meters— that's about 300 yards or 3 football fields. On a clear day, the view can extend for 67 km. Visitors can take elevators to the platforms or climb up the stairs: all 1,652 of them!

Accuracy

Accuracy has to do with both reasonableness and **rounding.**
The length of each **side** of the **square** below is measured
accurately to the nearest tenth of a meter. But the actual length
could be anywhere from 12.15 meters to 12.24 meters. (These
are the numbers that all round to 12.2.)

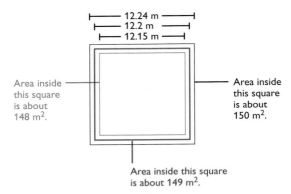

Since the side of the square could really be anywhere between
12.15 m and 12.24 m, the actual **area** may range anywhere
between 148 m^2 and 150 m^2. So is it reasonable to square the
side (12.2)^2 to get an area of 148.84 m^2? No, it isn't. Here is
why. The actual length is between 12.15 m and 12.24 m. The
area is between 148 m^2 and 150 m^2. Therefore 149 m^2 is
reasonable, but the last two digits in 148.84 are meaningless.

 Check It Out

3. The sides of a square measure 4.3 cm to the
   nearest tenth. Find the area of the square and
   discuss the accuracy of your answer.

4. A car travels approximately 178.5 mi in 4 hr. Is it
   accurate to say that its average speed is 44.625
   mi/hr? Why or why not?

8·1 EXERCISES

1. How many meters are there in a kilometer?
2. How many centimeters are there in a kilometer?
3. How many millimeters are there in 1 centimeter?

Which system of measurement uses the following?
4. ounces   5. liters
6. kilograms   7. yards
8. A car travels approximately 530 mi on 24 gal of gasoline. How many mi/gal does the car get? Explain your answer.

The measurements of each rectangle are given to the nearest tenth. Find the area of each figure. Explain your answers.

9.

10.4 m

4.6 m

10.

9.3 cm

8.3 cm

8·2 Length and Distance

About What Length?

When you get a feel for "about how long" or "around how far," it is easier to make estimations about length and distance. Here are some everyday items that will help you keep in mind what metric and customary units mean.

METRIC UNITS	CUSTOMARY UNITS
millimeter 1 mm about the width of the pin part of a push pin	inch 1 in. about the height of a regular postage stamp
centimeter 1 cm about the length of the pin part of a push pin	foot 1 ft a little longer than 2 CDs
meter 1 m a little more than the height of your desk	yard 1 yd. about the height of a stool

Check It Out

1. Use a customary rule or yardstick to measure common items. Find items that measure about an inch, a foot, and a yard.

2. Use a metric rule or meter stick to measure common items. Find items that measure about a millimeter, a centimeter, and a meter.

Metric and Customary Units

When you are calculating **length** and **distance,** you may encounter two different *systems of measurement* (p. 392). One is the metric system, and the other is the customary system.

Metric Equivalents

1 km	=	1,000 m	=	100,000 cm	=	1,000,000 mm
0.001 km	=	1 m	=	100 cm	=	1,000 mm
		0.01 m	=	1 cm	=	10 mm
		0.001 m	=	0.1 cm	=	1 mm

Customary Equivalents

1 mi	=	1,760 yd	=	5,280 ft	=	63,360 in.
$\frac{1}{1,760}$ mi	=	1 yd	=	3 ft	=	36 in.
		$\frac{1}{3}$ yd	=	1 ft	=	12 in.
		$\frac{1}{36}$ yd	=	$\frac{1}{12}$ ft	=	1 in.

CHANGING UNITS WITHIN A SYSTEM

How many feet are in $\frac{1}{8}$ mile?

units you have

1 mi = 5,280 ft

conversion factor
for new units

$\frac{1}{8} \times 5,280 = 660$

- Find the units you have where they equal 1 on the equivalents chart.

- Find the conversion factor.

- Multiply to get new units.

There are 660 ft in $\frac{1}{8}$ mi.

Check It Out

3. 2,200 m = ? km

4. 60 in. = ? ft

8.2 LENGTH AND DISTANCE

Conversions Between Systems

Once in a while, you may want to convert between the metric system and the customary system. You can use this conversion table to help.

CONVERSION TABLE

1 inch	=	25.4 millimeters	1 millimeter	=	0.0394 inch
1 inch	=	2.54 centimeters	1 centimeter	=	0.3937 inch
1 foot	=	0.3048 meter	1 meter	=	3.2808 feet
1 yard	=	0.914 meter	1 meter	=	1.0936 yards
1 mile	=	1.609 kilometers	1 kilometer	=	0.621 mile

To make a conversion, find the listing where the unit you have is 1. Multiply the number of units you have by the conversion factor for the new units.

Your friend in France says he can jump 127 cm. Should you be impressed? 1 cm = 0.3937 in. So 127 × 0.3937 = about 50 in. How far can you jump?

Most of the time you just need to estimate the conversion from one system to the other to get an idea of the size of your item. Round off numbers in the conversion table to simplify your thinking. Think that 1 meter is just a little more than 1 yard, 1 inch is between 2 and 3 centimeters, 1 mile is about $1\frac{1}{2}$ kilometers. So now when your friend in Senegal says she caught a fish 60 cm long, you know that it is between 20 and 30 in. long.

Check It Out

Make exact conversions. Use a calculator, and round to the nearest tenth.

5. Change 31 in. to cm.
6. Change 64 m to yd.
7. 5 km is about
   A. 3 mi,   B. 80 mi,   C. 8 mi
8. 100 ft is about
   A. 100 m,   B. 30 m,   C. 328 m

8·2 EXERCISES

Complete the conversions.

1. 1.93 km = ? m
2. 45 cm = ? mm
3. 750 cm = ? m
4. 820 m = ? km
5. 252 in. = ? ft
6. 10 yd = ? in.
7. 440 yd = ? mi
8. 1.5 mi = ? ft
9. 396 in. = ? yd
10. 12.5 ft = ? in.

Make exact conversions. Use a calculator and round to the nearest tenth.

11. Change 2 ft to cm.
12. Change 5.4 m to in.

13. Change 14 mi to km.
14. Change 420 mm to in.

15. Change 32 km to mi.
16. Change 15 yd to m.

Choose the conversion you estimate to be about right.

17. 550 yd is about
    A. 0.5 km   B. 5 km   C. 50 m
18. 1 m is about
    A. 1 ft   B. 10 ft C. 3 ft
19. 5 mi is about
    A. 10 km   B. 3 km   C. 8 km
20. 10 in. is about
    A. 25 cm   B. 4 cm   C. 250 cm

8·2 EXERCISES

8.3 Area, Volume, and Capacity

Area

Area is the measure of a surface. The walls in your room are surfaces. The large surface of the United States takes up an area of 3,787,319 sq mi. The area that the surface of a tire contacts on a wet road makes the difference between skidding and staying in control. Area is given in square units.

Area can be measured in metric units or customary units. Sometimes you might want to convert measurements within a measurement system. You can figure out the conversions by going back to the basic *dimensions* (p. 397). Below is a chart that provides the most common conversions.

Metric	Customary
$100 \text{ mm}^2 = 1 \text{ cm}^2$	$144 \text{ in.}^2 = 1 \text{ ft}^2$
$10{,}000 \text{ cm}^2 = 1 \text{ m}^2$	$9 \text{ ft}^2 = 1 \text{ yd}^2$
	$4{,}840 \text{ yd}^2 = 1 \text{ acre}$
	$640 \text{ acre} = 1 \text{ mi}^2$

To convert to a new unit, find the listing where the unit you have is one. Multiply the number of units you have by the conversion factor for the new unit. If the United States covers an area of about $3{,}800{,}000 \text{ mi}^2$, how many acres is it?

$1 \text{ mi}^2 = 640 \text{ acres}$,

so $3{,}800{,}000 \text{ mi}^2$ is $3{,}800{,}000 \times 640$ acres

$= 2{,}432{,}000{,}000$ acres

Check It Out

1. How many square feet are equal to 7 yd^2?
2. How many square centimeters are equal to 4 m^2?

Volume

Volume is expressed in cubic units. Here are the basic relationships among units of volume.

Metric	Customary
$1,000 \text{ mm}^3 = 1 \text{ cm}^3$	$1,728 \text{ in.}^3 = 1 \text{ ft}^3$
$1,000,000 \text{ cm}^3 = 1 \text{ m}^3$	$27 \text{ ft}^3 = 1 \text{ yd}^3$

CONVERTING VOLUME WITHIN A SYSTEM OF MEASUREMENT

Express the volume of the carton in cubic feet.

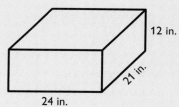

12 in.

21 in.

24 in.

$V = lwh$

$= 24 \times 21 \times 12$

$= 6,048 \text{ in.}^3$

$1,728 \text{ in.}^3 = 1 \text{ ft}^3$

$6,048 \div 1,728 = 3.5 \text{ ft}^3$

• Use a formula to find the volume (p. 60) using the units of the dimensions.

• Find the conversion factor.

• Multiply to convert to smaller units. Divide to convert to larger units.

So the volume of the carton is 3.5 ft^3.

• Include the unit of measurement in your answer.

Check It Out

3. Find the volume of a cube with sides of 50 cm. Convert your answer to m^3.

4. Find the volume of a rectangular prism with sides of 5 ft, 3 ft, and 18 ft. Convert your answer to cubic yards.

8·3 AREA, VOLUME, AND CAPACITY

Capacity

Capacity is closely related to volume, but there is a difference. A block of wood has volume but no capacity to hold liquid. The capacity of a container is a measure of the volume of liquid it will hold.

Metric	Customary
1 liter (L) = 1,000 milliliters (mL)	8 fl oz = 1 cup (c)
	2c = 1 pint (pt)
	2 pt = 1 quart (qt)
1 L = 1.057 qt	4 qt = 1 gallon (gal)

Note the use of *fl oz* (fluid ounce) in the table. This is to distinguish it from *oz* (ounce) which is a unit of weight (16 oz = 1 lb). Fluid ounce is a unit of capacity (16 fl oz = 1 pint). There is a connection between ounce and fluid ounce. A pint of water weighs about a pound, so a fluid ounce of water weighs about an ounce. For water, as well as for most other liquids, *fluid ounce* and *ounce* are equivalent, and the "fl" is sometimes omitted (for example, "8 oz = 1 cup"). To be correct, though, use *ounce* for weight only and *fluid ounce* for capacity. For liquids that weigh considerably more or less than water, the difference is significant.

A gallon of juice costs $4.88. How much does it cost per liter?

Figure out how many liters are in a gallon. There are 4 quarts in a gallon, so there are 4 × 1.057 liters, or 4.228 liters in a gallon. So 1 liter of juice costs $4.88 ÷ 4.228 or $1.15.

Check It Out

5. Which is the better buy: 1 liter of milk for $1.04 or 1 quart of milk for $0.95?

8·3 EXERCISES

Identify each unit as a measure of distance, area, volume, or capacity.
 1. liter
 2. km^3
 3. mm
 4. mi^2

Give the area of the rectangle in each of the units.
 5. $in.^2$
 6. ft^2

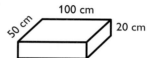

20 in.

18 in.

Give the volume of the rectangular prism in each of the units.
 7. cm^3
 8. m^3

100 cm

50 cm

20 cm

 9. Find the area of a square with sides of 4.5 ft. Give the answer in square yards.
10. Find the volume of a cube with sides of 150 cm. Give the answer in cubic meters.

Complete the conversions.
11. 20 pt = ? gal
12. 12 c = ? qt
13. 1 qt = ? fl oz
14. 3.5 gal = ? c
15. 96 fl oz = ? gal
16. 2.75 gal = ? pt
17. 400 mL = ? L
18. 3 L = ? mL
19. 1,750 mL = ? L

20. Box A has a volume of 140 ft^3, and box B has a volume of 5 yd^3. Which is the larger box?

8·3 EXERCISES

8·4 Mass and Weight

Technically, mass and weight are different. Mass is the amount of substance you have. Weight is the pull of gravity on the amount of substance. On Earth, mass and weight are equal at sea level and about equal at other elevations. But on the moon, mass and weight can be quite different.

Your mass would be the same on the moon as it is here on Earth. But, if you weigh 100 pounds on Earth, you would weigh about $16\frac{2}{3}$ pounds on the moon. That is because the gravitational pull of the moon is only $\frac{1}{6}$ that of the Earth.

The customary system measures weight. The metric system measures mass.

Metric Equivalents

1 kg = 1,000 g = 1,000,000 mg
0.001 kg = 1 g = 1,000 mg
0.000001 kg = 0.001 g = 1 mg

Customary Equivalents

1 T = 2,000 lb = 32,000 oz
0.0005 T = 1 lb = 16 oz
0.0625 lb = 1 oz

$$1 \text{ lb} \approx 0.4536 \text{ kg}$$
$$1 \text{ kg} \approx 2.205 \text{ lb}$$

To convert from one unit of mass or weight to another, first find the 1 for the units you have in the equivalents chart. Then multiply the number of units you have by the conversion factor for the new units.

If you have 64 oz of peanut butter, how many pounds do you have? 1 oz = 0.0625 lb, so 64 oz = 64 × 0.0625 lb = 4 lb. You have 4 lb of peanut butter.

Check It Out

Complete the following conversions.

1. 2.1 T = ? lb
2. 640 mg = ? g
3. 40 kg = ? lb

8·4 EXERCISES

Complete the conversions.
1. 40 oz = ? lb
2. 3,500 lb = ? T
3. 12.35 g = ? mg
4. 6,040 mg = ? g
5. 4.5 lb = ? oz
6. 85 g = ? kg
7. 3,200 oz = ? T
8. 150,000 mg = ? kg
9. 500 oz = ? lb
10. 45 mg = ? g
11. 0.02 kg = ? mg
12. 0.5 T = ? oz
13. 1 T = ? kg
14. 100 g = ? oz
15. 10 kg = ? lb
16. 8 oz = ? g
17. 15 lb = ? kg

18. A recipe calls for 18 oz of flour for 6 people. How many pounds of flour do you need to make the recipe for 24 people?
19. A 2-lb bag of dried fruit costs $6.40. A 5-oz box of the same fruit costs $1.20. Which is the better deal?
20. How many 4-oz balls of wool would you have to buy if you needed 2 kg of wool?

Poor SID

SID is a crash-test dummy. After a crash, SID goes to the laboratory for a readjustment of sensors and perhaps a replacement head or other body parts. Because of the forces at work when a car crashes, body parts weigh as much as 20 times their normal weight.

The weight of a body changes during a crash. Does the mass of the body also change? See Hot Solutions for the answer.

8·5 Time

Time measures the interval between two or more events. You can measure time with a very short unit—a second—a very long unit—a millennium—and many units in between.

> 1,000,000 seconds before 12:00 A.M., January 1, 2000 is 10:13:20 A.M., December 20, 1999. 1,000,000 hours before 12:00 A.M., January 1, 2000 is 8 A.M., December 8, 1885.

60 seconds (sec) = 1 minute (min)	365 da = 1 year (yr)
60 min = 1 hour (hr)	10 yr = 1 decade
24 hr = 1 day (da)	100 yr = 1 century
7 da = 1 week (wk)	1,000 yr = 1 millennium

Working with Time Units

Like other kinds of measurement, you can convert one unit of time to another by using the information in the table above.

Hyeran is exactly 12 years old. Her age in months is 12×12, or 144 months.

Leap Years

Every four years, February has an extra day. These 366 day years are called *leap years*. Leap years are divisible by 4, but not by 100. However, years that are divisible by 400 are leap years. The year 1996 is a leap year, but the year 1900 is not. The year 2000 is also a leap year.

Check It Out

1. How many minutes are there in a leap year?
2. It is Monday morning, at 5 min past midnight. What day of the week will it be in 1,000,000 seconds time?

8·5 EXERCISES

Complete the conversions.
1. 144 hr = ? da
2. 9 wk = ? da
3. 1 da = ? sec
4. 2 yr = ? wk
5. 1 wk = ? hr
6. 1 millenium = ? centuries

7. If you can walk $\frac{1}{2}$ mi in 10 minutes, what is your speed in miles/hour?
8. About how many hours are there in a century?
9. A 12-hour clock chimes once at one o'clock, twice at two o'clock, and so on. How many times does it chime in all in one week?

10. Sound travels at about 1,088 ft/sec at sea level. What is the approximate speed of sound in miles/hour?

The World's Largest Reptile

Would it surprise you to learn that the world's largest reptile is a turtle? The leatherback turtle can weigh as much as 2,000 pounds. By comparison, an adult male crocodile weighs about 1,000 pounds.

The leatherback has existed in its current form for over 20 million years, but this prehistoric giant is now endangered. If after 20 million years of existence, the leatherback was to become extinct, how many times longer than Homo sapiens will it have existed? Assume Homo sapiens has been around 4,000 millennia. See Hot Solutions for answer.

8·6 Size and Scale

Similar Figures

Similar figures are figures that have exactly the same shape. When two figures are similar, one may be larger than the other.

DECIDING IF TWO FIGURES ARE SIMILAR

Are these two rectangles similar?

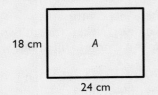
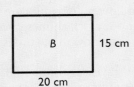

18 cm   A

B   15 cm

24 cm   20 cm

$$\frac{24}{20} \overset{?}{=} \frac{18}{15}$$

- Set up the **ratios:** $\frac{\text{length } A}{\text{length } B} \overset{?}{=} \frac{\text{width } A}{\text{width } B}$

$$15 \times 24 \overset{?}{=} 18 \times 20$$
$$360 = 360$$

- Cross multiply to see if ratios are equal.

So the rectangles are similar.

- If all sides have equal ratios, the figures are similar.

Check It Out

1. Which figures are similar?

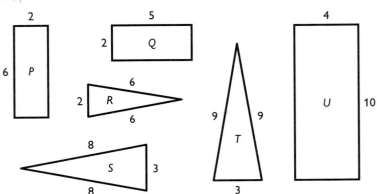

2

6   P

5

2   Q

6

2   R

6

8

S   3

8

9   9

T

3

4

U   10

Scale Factors

A **scale factor** indicates the ratio of sizes of two similar figures.

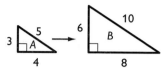

Triangle A is similar to triangle B. $\triangle B$ is two times larger than $\triangle A$. The scale factor is 2.

FINDING THE SCALE FACTOR

What is the scale factor for these similar rectangles?

J ⎯⎯⎯ 21 ⎯⎯⎯ K	J' ⎯⎯ 15 ⎯⎯ K'
7	5
M ⎯⎯⎯⎯⎯ L	M' ⎯⎯⎯ L'

$\frac{J'K'}{JK} = \frac{15}{21}$

- Decide which figure is the "original figure."
- Make a ratio of corresponding sides:

$$\frac{\text{new figure}}{\text{original figure}}$$

$= \frac{5}{7}$

- Reduce, if possible.

The scale factor of the two pentagons is $\frac{5}{7}$.

When a figure is enlarged, the scale factor is greater than one. When two similar figures are identical in size, the scale factor is equal to one. When a figure is reduced, the scale factor is less than one.

Check It Out

Find the scale factors.

2.

6 ⎯ 4 ▢ → 10 ⎓ 15 ▭

3. 8 ◹ 8 / 12 → 6 ◹ 6 / 9

8·6 SIZE AND SCALE

Scale Factors and Area

Scale factor refers to a ratio of lengths only, not of areas.

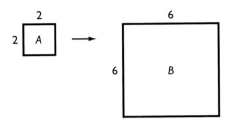

For the squares, the scale factor is 3 because the ratio of sides is $\frac{6}{2} = 3$. Notice that, while the scale factor is 3, the ratio of the areas is 9.

$$\frac{\text{Area of } B}{\text{Area of } A} = \frac{6^2}{2^2} = \frac{36}{4} = 9$$

The scale factor is $\frac{1}{2}$. What is the ratio of the areas?

$$\frac{\text{Area of } D}{\text{Area of } C} = \frac{3 \times 2}{6 \times 4} = \frac{6}{24} = \frac{1}{4}$$

The ratio of the areas is $\frac{1}{4}$.

In general, the ratio of the areas of two similar figures is the *square* of the scale factor.

Check It Out

4. The scale factor for two similar figures is $\frac{5}{4}$. What is the ratio of the areas?
5. The scale of a house model is 1 ft = 10 ft. How much area of the house does an area of 1 ft^2 on the model represent?

8·6 EXERCISES

Give the scale factor.

1.

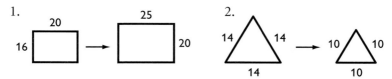

2.

3. A document measuring 28 cm by 20 cm is enlarged by a scale factor of $\frac{3}{2}$. What are the dimensions of the enlarged document?
4. A photograph 8 in. long and 6 in. wide is reduced. The reduced photo is 6 in. long. How wide is it?
5. The scale factor of the two similar triangles is $\frac{1}{2}$. What are the dimensions of the smaller triangle?

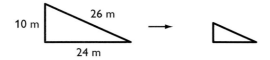

6. The scale of a map is 1 in. = 10 mi. On the map, two towns are 3.5 in. apart. How far apart are they actually?
7. The scale of a map is 1 cm = 5 km. What length on the map would a road be that is 12.5 km long?
8. The scale factor of two similar figures is 4. If the smaller rectangle has an area of 2 ft^2, what is the area of the larger figure?
9. A square park has sides of 12 km. What is its area on a map that has a scale of 1 cm = 10 km?
10. The scale factor of the two similar pentagons is $\frac{2}{3}$. Find the lengths of r, s, and t.

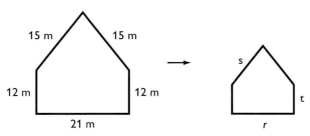

What have you learned?

You can use the problems and list of words below to see what you have learned in this chapter. You can find out more about a particular problem or word by referring to the boldfaced topic number (for example, **8•2**).

Problem Set

1. Name one metric and one customary unit of capacity.   **8•1**
2. Which system of measurement is based on powers of ten?   **8•1**
3. Complete the equivalent measures.   **8•1**
   1 km = ? m = ? cm = ? mm

Complete the conversions.   **8•2**

4. 32.65 km = ? m
5. 1.28 km = ? mm
6. $5\frac{1}{2}$ yd = ? in.
7. 7,920 ft = ? mi

Use the rectangle to answer items 8–11.

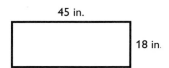

45 in.

18 in.

8. What is the figure's perimeter in feet?   **8•2**
9. What is the figure's perimeter in yards?   **8•2**
10. What is the figure's area in square feet?   **8•3**
11. What is the figure's area in square yards?   **8•3**

Complete the conversions.   **8•3**

12. 10.04 cm^2 = ? mm^2
13. 11 yd^2 = ? ft^2
14. 4 yd^2 = ? in.^2
15. 7.113 m^3 = ? cm^3
16. 135 ft^3 = ? yd^3

Use the rectangular prism to answer items 17–19. **8•3**

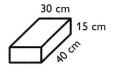

30 cm

15 cm

40 cm

17. What is the prism's volume in cubic millimeters?
18. What is the prism's volume in cubic meters?
19. If 1 in. = 2.54 cm, what is the approximate volume of the prism in cubic inches?

20. A recipe calls for 14 oz of rice for 6 people . How many pounds of rice do you need to make the recipe for 36 people? **8•4**
21. How many 175-g packets of flour would you have to buy if you needed at least 4.4 kg of flour? **8•4**
22. How many days does 1,000,000 sec take? **8•5**
23. About how many minutes are there in a century? **8•5**

A photograph that was 5 in. long by 3 in. wide was enlarged so that the new photo was 1 ft long. **8•6**
24. What is the new width of the photo?
25. If the area of the original photo is x in.^{2}, what is the area of the new photo in terms of x?

---

hot words

WRITE DEFINITIONS FOR THE FOLLOWING WORDS.

accuracy **8•1**
area **8•1**
customary system **8•1**
distance **8•2**

factors **8•1**
fractions **8•1**
length **8•2**
metric system **8•1**
perimeter **8•1**
power **8•1**
ratio **8•6**
rectangle **8•1**

rounding **8•1**
scale factor **8•6**
side **8•1**
similar figures **8•1**
square **8•1**
time **8•5**
volume **8•3**

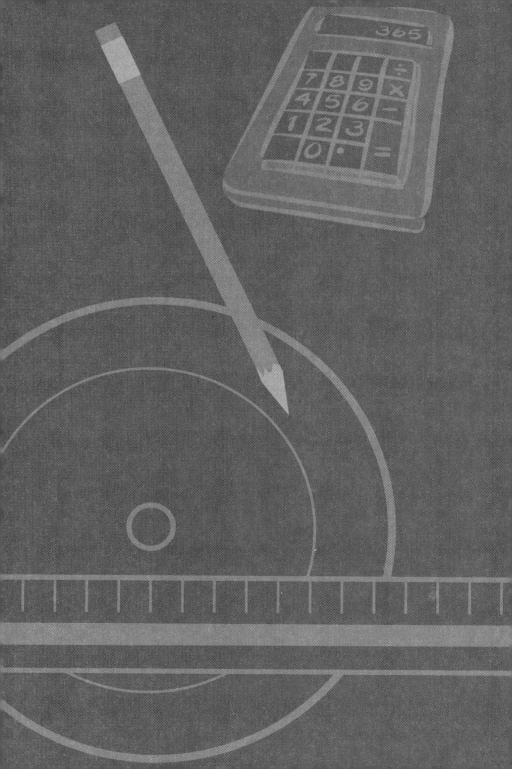

Tools

What do you already know?

You can use the problems and list of words below to see what you already know about this chapter. The answers to the problems are in Hot Solutions at the back of the book, and the definitions of the words are in Hot Words at the front of the book. You can find out more about a particular problem or word by referring to the boldfaced topic number (for example, **1•2**).

Problem Set

Use your calculator for items 1–6. **9•1**

1. $55 + 8 \times 12$
2. 150% of 1,700

Round answers to the nearest tenth.

3. $14 + 4.25 \times (-5) + 29$
4. $-80 + 28 \div 2.5 - 8.75$

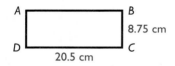

A   B
8.75 cm
D   C
20.5 cm

5. Find the perimeter of rectangle *ABCD*.
6. Find the area of rectangle *ABCD*.

Use a scientific calculator for items 7–12. Round decimal answers to the nearest hundredth. **9•2**

7. 6.4^3
8. Find the reciprocal of 5.3.
9. Find the square of 8.2.
10. Find the square root of 8.2.
11. $(7 \times 10^3) \times (2 \times 10^5)$
12. $1.4 \times (25 \times 2.4)$

13. What is the measure of ∠*VRS*? **9•3**
14. What is the measure of ∠*SRT*? **9•3**
15. What is the measure of ∠*TRV*? **9•3**
16. Does $\overrightarrow{RT}$ divide ∠*SRV* into two equal angles? **9•3**

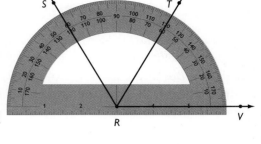

17. What basic construction tool would you use to draw a circle? **9•3**

For items 18–20, refer to the spreadsheet.

18. Name the cell holding 40. **9•4**
19. A formula for cell C2 is C1 + 100. Name another formula for cell C2. **9•4**
20. Cell D1 contains the number 2,000 and no formula. After using the command fill down, what number will be in cell D6? **9•4**

File   Edit

Fill down
Fill right

	A	B	C	D
1	5	20	100	2000
2	5	40	200	
3	5	60	300	
4				

CHAPTER 9

*hot***words**

	distance **9•3**	radius **9•1**
	factorial **9•2**	ray **9•3**
	formula **9•4**	reciprocal **9•2**
angle **9•2**	horizontal **9•4**	root **9•2**
arc **9•3**	negative number	row **9•4**
cell **9•4**	**9•1**	spreadsheet **9•4**
circle **9•1**	parentheses **9•2**	square **9•2**
column **9•4**	percent **9•1**	square root **9•1**
cube **9•2**	perimeter **9•4**	tangent **9•2**
cube root **9•2**	pi **9•1**	vertex **9•3**
decimal **9•1**	point **9•3**	vertical **9•4**
degree **9•2**	power **9•2**	

WHAT DO YOU KNOW?

9·1 Four-Function Calculator

People use calculators to make mathematical tasks easier. You might have seen your parents balance their checkbooks using a calculator. But a calculator is not always the fastest way to do a mathematical task. If your answer does not need to be exact, it might be faster to estimate. Sometimes you can do the problem in your head quickly, or a pencil and paper might be a better method. Calculators are particularly helpful for problems with many numbers or with numbers that have many digits.

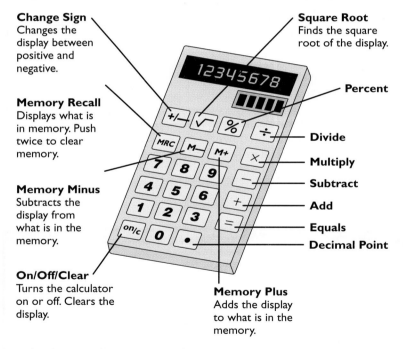

Change Sign
Changes the display between positive and negative.

Square Root
Finds the square root of the display.

Percent

Memory Recall
Displays what is in memory. Push twice to clear memory.

Divide

Multiply

Memory Minus
Subtracts the display from what is in the memory.

Subtract

Add

Equals

Decimal Point

On/Off/Clear
Turns the calculator on or off. Clears the display.

Memory Plus
Adds the display to what is in the memory.

A calculator only gives you the answer to the problem you enter. Always have an estimate of the answer you expect. Then you can compare the calculator answer to your estimate to be sure you entered the problem correctly.

Basic Operations

Adding, subtracting, multiplying, and dividing are fairly straightforward.

Operation	Problem	Calculator Keys	Display
Addition	$40 + 15.8$	40 ⊞ 15.8 ⊟	55.8
Subtraction	$18 - 23$	18 ⊟ 23 ⊟	−5.
Multiplication	12.5×3.3	12.5 ⊠ 3.3 ⊟	41.25
Division	$8 \div 20$	8 ⊟ 20 ⊟	0.4

Negative Numbers

To enter a **negative number** into your calculator, you press ⌊+/−⌋ after you enter the number.

Problem	Calculator Keys	Display
$-22 + 17$	22 ⌊+/−⌋ ⊞ 17 ⊟	−5.
$30 - (-4.5)$	30 ⊟ 4.5 ⌊+/−⌋ ⊟	34.5
$20 \times (-8)$	20 ⊠ 8 ⌊+/−⌋ ⊟	−160.
$-10 \div (-2)$	10 ⌊+/−⌋ ⊟ 2 ⌊+/−⌋ ⊟	5.

Check It Out

Find each answer on a calculator.
1. $19.5 + 7.2$
2. $31.8 - 23.9$
3. $10 \times (-0.5)$
4. $-24 \div 0.6$

9.1 FOUR-FUNCTION CALCULATOR

Memory

For complex or multi-step problems, use memory. You operate memory with three special keys. Here is the way many calculators operate. If yours does not work this way, check the instructions which came with your calculator.

Key **Function**

MRC — One push displays (recalls) what is in memory. Push twice to clear memory.

M+ — Adds display to what is in memory.

M− — Subtracts display from what is in memory.

When calculator memory contains something other than zero, the display will show M⎕ along with whatever number the display currently shows. What you do on your calculator does not change memory unless you use the special memory keys.

To solve $20 + 65 + 46 \times 5 + 50 - 5^2$, you could use the following keystrokes with your calculator:

Keystrokes	Display
MRC MRC C	0.
5 × 5 M−	M 25.
46 × 5 M+	M 230.
20 + 65 M+	M 85.
50 M+	M 50.
MRC	M 340.

Your answer is 340. Notice the use of *order of operations* (p. 80).

Check It Out

Use memory to find each answer.

5. $7 \times 6 - 13 \times 2 + 12^2$
6. $-10 + 5^3 - (-6) \times 40$
7. $8^4 \times 2 + (-15) \times 10 + (-21)$
8. $15^2 + 22 \times (-4) - (-80)$

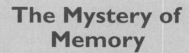

The Mystery of Memory

Researchers distinguish three layers of factual memory. Immediate memory lasts a couple of seconds and holds information long enough to perform a rapid task, such as looking up and dialing a phone number. Short-term memory retains and recalls data for a few seconds to a few minutes. Long-term memory stores information for a few months or a life-time.

Play this memory-calculator game with a friend. Turn on a four-function calculator and clear its memory. Take turns entering numbers less than 50 and pressing the M+ key. When one player thinks the total in the calculator's memory is 200 or greater, check by pressing the MR key. Were you able to add the numbers correctly in your head?

Which layer or layers of memory do you think you used to play this game? See Hot Solutions for answer.

9·1 FOUR-FUNCTION CALCULATOR

Special Keys

Some calculators have keys with special functions to save time.

Key	Function
$\sqrt{x}$	Finds the **square root** of the display.
%	Changes display to the **decimal** expression of a **percent.**
π	Automatically enters **pi** to as many places as your calculator holds.

The $\boxed{\%}$ and $\boxed{\pi}$ keys save you time by saving you keystrokes. The $\boxed{\sqrt{}}$ key allows you to find square roots precisely, something difficult to do by hand. See how they work in the examples below.

Problem: $18 + \sqrt{225}$

Keystrokes: $18 \boxed{+} 225 \boxed{\sqrt{}} \boxed{=}$

Final display: $\boxed{\qquad 33.}$

If you try to take a square root of a negative number, your calculator will display an error message, such as:

$25 \boxed{+/-} \boxed{\sqrt{}} \boxed{\text{E} \qquad 5.}$.

There is no square root of -25, because no number times itself can give a negative number.

Problem: Find 25% of 50.

Keystrokes: $50 \boxed{\times} 25 \boxed{\%}$

Final display: $\boxed{\qquad 12.5}$

The $\boxed{\%}$ key only changes a percent to its decimal form. If you know how to convert percents to decimals, you probably will not use the $\boxed{\%}$ key much.

Problem: Find the area of a **circle** with **radius** 4.
(Use formula $A = \pi r^2$.)

Keystrokes: $\boxed{\pi} \boxed{\times} 4 \boxed{\times} 4 \boxed{=}$

Final display: $\boxed{\qquad 50.27}$

If your calculator does not have the $\boxed{\pi}$ key, you can use 3.14 or 3.1416 as an approximation for π.

 Check It Out

9. Without using the calculator, tell what the display would be if you entered: $10 \boxed{\text{M+}} 5 \boxed{\times} 4 \boxed{+} \boxed{\text{MRC}} \boxed{=}$.

10. Use memory functions to find the answer to $220 - 6^2 \times (-10)$.

11. Find the square root of 484.

12. Find 35% of 250.

9·1 EXERCISES

Find the value of each expression using your calculator.

1. $26.8 + 43.3$
2. $65.82 - 30.7$
3. $-16.5 - 7.46$

4. $15 \times 32 \times 10$
5. $-8 + 40 \times (-6)$
6. $75 - 19 \times 20$

7. $\sqrt{144} - 5$
8. $64 + \sqrt{256}$
9. $72 \div 16 + 12$

10. $8 \div (-40)$
11. 15% of 180
12. 150% of 240

13. $128 - \sqrt{169}$
14. $\sqrt{900} \div 4.8 + 12.75$
15. $8 \times \sqrt{24} + 11$

Use a calculator for items 16–19.

16. Find the area if $x = 3.5$ cm.
17. Find the perimeter if $x = 2.48$ cm.

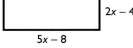

18. Find the area if $a = 5$.
19. Find the circumference if $a = 1$.

20. Find the area of $\triangle RQP$.
21. Find the perimeter of $\triangle RQP$.
    (*Remember:* $a^2 + b^2 = c^2$.)
22. Find the circumference of circle Q.
23. Find the area of circle Q.
24. Find the length of line segment RP.
25. Find the shaded area of circle Q.

9•2 Scientific Calculator

Every mathematician and scientist has a scientific calculator to help quickly and accurately solve complex equations. Scientific calculators vary widely, some with a few functions and others with many functions. Some calculators can even be programmed with functions of your choosing. The calculator below shows functions you might find on your scientific calculator.

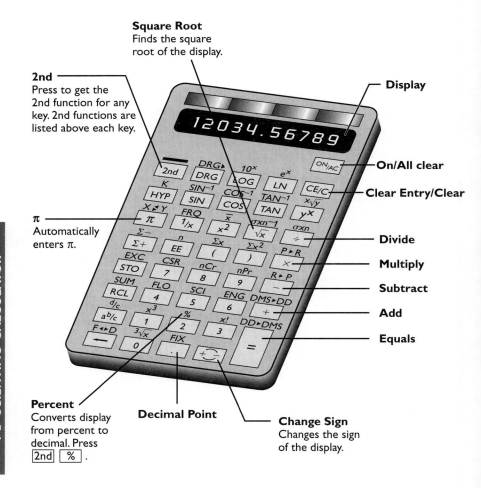

Square Root
Finds the square root of the display.

2nd
Press to get the 2nd function for any key. 2nd functions are listed above each key.

Display

π
Automatically enters π.

On/All clear

Clear Entry/Clear

Divide

Multiply

Subtract

Add

Equals

Percent
Converts display from percent to decimal. Press [2nd] [%].

Decimal Point

Change Sign
Changes the sign of the display.

Frequently Used Functions

Since each scientific calculator is set up differently, your calculator may not work exactly as below. These keystrokes work with the calculator illustrated on page 424. Use the reference book or card that came with your calculator to perform similar functions. See the index to find more information about the mathematics here.

Function	Problem	Keystrokes
Cube Root $\boxed{\sqrt[3]{x}}$ Finds the cube root of the display.	$\sqrt[3]{125}$	125 $\boxed{\text{2nd}}$ $\boxed{\sqrt[3]{x}}$ $\boxed{\qquad 5.}$
Cube $\boxed{x^3}$ Finds the cube of the display.	12^3	12 $\boxed{\text{2nd}}$ $\boxed{x^3}$ $\boxed{\qquad 1728.}$
Factorial $\boxed{x!}$ Finds the factorial of the display.	$4!$	4 $\boxed{\text{2nd}}$ $\boxed{x!}$ $\boxed{\qquad 24.}$
Fix number of **decimal places**. $\boxed{\text{FIX}}$ Rounds display to number of places you determine.	Round 2.189 to the hundredths place.	2.189 $\boxed{\text{2nd}}$ $\boxed{\text{FIX}}$ 2 $\boxed{\qquad 2.19}$
Parentheses $\boxed{(}\boxed{)}$ Use to group calculations.	$(3 + 5) \times 11$	11 $\boxed{\times}$ $\boxed{(}$ 3 $\boxed{+}$ 5 $\boxed{)}$ $\boxed{=}$ $\boxed{\qquad 88.}$
Powers $\boxed{y^x}$ Finds the x power of the display.	21^4	21 $\boxed{y^x}$ 4 $\boxed{=}$ $\boxed{\qquad 194481.}$
Powers of ten $\boxed{10^x}$ Raises ten to the power displayed.	10^4	4 $\boxed{\text{2nd}}$ $\boxed{10^x}$ $\boxed{\qquad 10000.}$

9-2 SCIENTIFIC CALCULATOR

Function	Problem	Keystrokes
Reciprocal $\boxed{1/x}$ Finds the reciprocal of the display.	Find the reciprocal of 5.	$5 \ \boxed{1/x} \ \boxed{\qquad 0.2}$
Roots $\boxed{\sqrt[x]{y}}$ Finds the x root of the display.	$\sqrt[5]{7,776}$	$7776 \ \boxed{\text{2nd}} \ \boxed{\sqrt[x]{y}} \ 5 \ \boxed{=}$ $\boxed{\qquad 6.}$
Square $\boxed{x^2}$ Finds the square of the display.	13^2	$13 \ \boxed{x^2} \ \boxed{\qquad 169.}$

 Check It Out
Use your calculator to find the following.
1. 8!   2. 11^4
Use your calculator to find the following to the nearest thousandth.
3. the reciprocal of 8
4. $(9^2 - 11^4 + \sqrt[3]{1728}) \div 4$

Tangent

Use $\boxed{\text{TAN}}$ to find the **tangent** of an **angle** (p. 382). Angles can be expressed as **degrees** or radians. Use the "Degree" or "DRG" or "DR" key to put your calculator in *degree* mode.

Find x to the nearest tenth.

$\tan 42° = \frac{x}{5}$ $x = 5 \times \tan 42°$

To find x, enter $5 \ \boxed{\times} \ 42 \ \boxed{\text{TAN}} \ \boxed{=} \ 4.5020202$
$x \approx 4.5$.

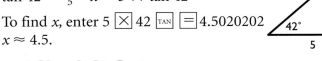

 Check It Out
5. What is the tangent of 58° expressed to the nearest tenth?
6. What is the tangent of 43° expressed to the nearest hundredth?

9·2 EXERCISES

Use a scientific calculator to find the following.
1. 45^2 2. 20^3 3. 7^4 4. 1.5^4

For answers with decimals, give your answer to the nearest hundredth.
5. $\frac{33}{\pi}$ 6. $7(\pi)$ 7. $\frac{1}{4}$ 8. $\frac{3}{\pi}$
9. $(12 + 2.6)^2 + 8$ 10. $64 - (12 \div 3.5)$
11. $3! \times 5!$ 12. $6! \div 4!$ 13. $8! + 7!$
14. 10^{-2} 15. $\sqrt[3]{5,832}$ 16. reciprocal of 20

Give the tangent of the angle to the nearest hundredth.
17. $48°$ 18. $73°$ 19. $45°$

20. At a distance 64 ft from a building, the angle of the line of sight to the top of the building is $32°$. How tall is the building to the nearest foot?

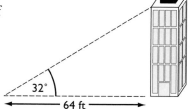

64 ft

Magic Numbers

Ordinary calculations with integers sometimes result in astonishing patterns.

$1 \times 8 + 1 = 9$

$12 \times 8 + 2 = 98$

$123 \times 8 + 3 = 987$

$1234 \times 8 + 4 = 9876$

Write the equation that comes next in this pattern. Check your answer with a calculator.

9.3 Geometry Tools

The Ruler

If you need to measure the dimensions of an object, or if you need to measure reasonably short **distances,** use a ruler.

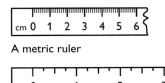

A metric ruler

A customary ruler

To get an accurate measure, be sure one end of the item being measured lines up with zero on your ruler.

The pencil below is measured first to the nearest tenth of a centimeter and then to the nearest eighth of an inch.

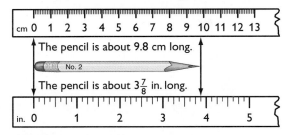

The pencil is about 9.8 cm long.

The pencil is about $3\frac{7}{8}$ in. long.

Check It Out

Use your ruler. Measure each line segment to the nearest tenth of a centimeter or the nearest eighth of an inch.

1. •————————————————•

2. •————————————————•

3. •————————————•

4. •——————————•

The Protractor

Measure angles with a *protractor.* There are many different protractors.

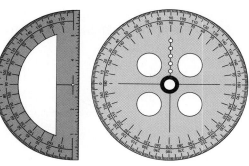

The key is to find the point on each protractor to which you align the **vertex** of the angle.

MEASURING ANGLES WITH A PROTRACTOR

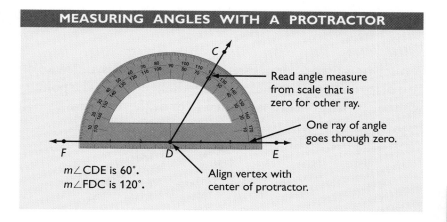

Read angle measure from scale that is zero for other ray.

One ray of angle goes through zero.

m∠CDE is 60°.
m∠FDC is 120°.

Align vertex with center of protractor.

To draw an angle with a given measure, draw one **ray** first and position the center of the protractor at the endpoint. Then make a dot at the desired measure (45°, in this example).

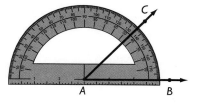

Connect *A* and *C*. Then ∠*BAC* is a 45° angle.

Check It Out

Measure each angle to the nearest degree using your protractor.

5.

6.

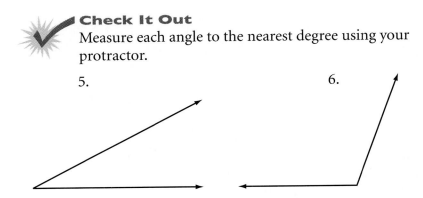

Compass

A *compass* is used to draw circles or parts of circles, called **arcs**. You place one **point** at the center and hold it there. The point with the pencil attached is pivoted to draw the arc or circle.

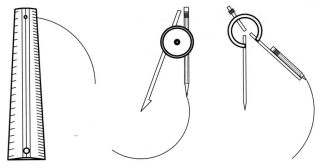

The distance between the point that is stationary (the center) and the pencil is the radius. Some compasses allow you to set the radius exactly.

For a review of *circles*, see page 372.

To draw a circle with a radius of $1\frac{1}{2}$ inches, set the distance between the stationary point of your compass and the pencil at $1\frac{1}{2}$ inches. Draw a circle.

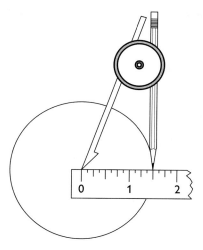

 Check It Out

Draw circles with the following radius measures.

7. radius 3 in. or 7.6 cm
8. radius 6 cm or 2.4 in.
9. radius 4 cm or 1.6 in.
10. radius 4 in. or 10.2 cm

Construction Problem

A construction is a drawing problem in geometry that permits the use of only the straightedge and the compass. When you make a construction using the straightedge and the compass, you have to use what you know about geometry.

CONSTRUCTING AN ANGLE BISECTOR

Given ∠A, construct the angle *bisector*, or a line that divides the angle exactly in half.

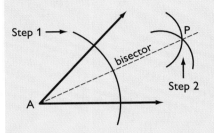

Step 1 →
bisector
P
Step 2
A

- Using a compass, draw an arc centered at A that intersects both sides of ∠A.
- Draw two more arcs centered where the first arc intersects the sides. Call the point where these arcs intersect point P.
- Connect A and P.

Ray AP divides ∠A into two equal angles, each being half the measure of ∠A. Ray AP is the angle bisector.

Check It Out

11. If you bisect an angle of 44°, what is the measure of the two angles formed by the angle bisector?

12. Draw a 50° angle. Then construct the angle bisector.

13. What is the measure of an angle bisecting a right angle?

14. How could you use a protractor to check to see if you correctly bisected an angle? Explain.

Ferns, Fractals, and Branches

Have you ever looked closely at a fern frond? Did you notice that the frond is made up of smaller and smaller parts that look self-similar? That is, the smaller parts of the frond look very much like the frond itself.

In mathematics, shapes that have complex and detailed structures at any level of magnification are called *fractals*. Like ferns, many natural objects display fractal-like patterns. But they are not fractals in the mathematical sense, because their complexity does not go on forever.

Use any of your geometric tools to draw a fractal pattern. It can be a design based on a geometric shape, or it can be a pattern like one found in nature.

9.3 EXERCISES

Using a ruler, measure the length of each side of $\triangle ABC$. Give your answer in inches or centimeters, rounded to the nearest $\frac{1}{8}$ in. or $\frac{1}{10}$ cm.

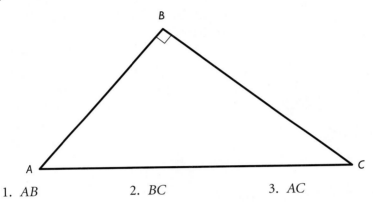

1. AB 2. BC 3. AC

Using a protractor, measure each angle in $\triangle ABC$.

4. $\angle A$ 5. $\angle B$ 6. $\angle C$

7. If all of the sides of a triangle are equal, what do you know about the measure of each of the angles? Explain.

Match each tool with the function.

Tool	Function
8. ruler	A. measures angles
9. compass	B. measures distance
10. protractor	C. draws circles or arcs

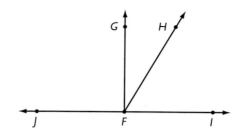

Write the measures of:
11. ∠GFH
12. ∠HFJ
13. ∠IFJ
14. ∠HFI
15. ∠JFG

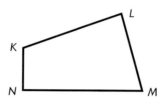

Write the measures of:
16. ∠NML
17. ∠MLK
18. ∠KNM
19. ∠LKN

20. Use a protractor to copy ∠MLK.
21. Use a compass and straightedge to bisect the angle you drew for item 20 above.

Using a ruler, protractor, and compass, copy the figures below.

22.   23.   24.   25.

9.4 Spreadsheets

What Is a Spreadsheet?

People have used **spreadsheets** as a tool to keep track of information, such as finances, for a long time. Spreadsheets were a paper-and-pencil math tool before becoming computerized. You may be familiar with computer spreadsheet programs.

A spreadsheet is a computer tool where information is arranged into **cells** within a grid and calculations are performed within the cells. When one cell is changed, all other cells that depend on it automatically change.

Spreadsheets are organized into **rows** and **columns.** Rows are **horizontal** and are numbered. Columns are **vertical** and are named by capital letters. The cells are named for their rows and columns.

File Edit

	A	B	C	D
1	1	3	1	
2	2	6	4	
3	3	9	9	
4	4	12	16	
5	5	15	25	
6				
7				
8				

The cell A3 is in Column A, Row 3. In this spreadsheet, there is a 3 in cell A3.

Check It Out

In the spreadsheet above, what number appears in each cell?

1. A1   2. B3   3. C4

Answer the next questions true or false.

4. A column is labeled with numbers.

5. A row is vertical.

Spreadsheet Formulas

A cell can contain a number, or it may contain the information it needs to generate a number. A **formula** generates a number dependent on other cells in the spreadsheet. The way the formulas are written depends on the particular spreadsheet computer software you are using. You enter a formula and the value generated shows, not the formula.

CREATING A SPREADSHEET FORMULA

	A	B	C	D
1	Item	Price	Qty	Total
2	sweater	$30	2	$60
3	pants	$18	2	
4	shirt	$10	4	
5				
6				

Express the value of the cell in relationship to other cells.

Total = Price × Qty
D2 = B2 × C2

If you change the value of a cell and a formula depends on it, the result of the formula will change.

In the spreadsheet above, if you entered 3 sweaters instead of 2 (C2 = 3), the Total column would automatically change to $90.

Check It Out

Use the spreadsheet above. If the Total is always figured the same way, write the formula for:
6. D3
7. D4
8. What is the total spent on shirts?
9. What is the total spent on pants?

Fill Down and Fill Right

Now that you know the basics, let's look at some ways to make spreadsheets do even more of the work for you. *Fill down* and *fill right* are two spreadsheet commands that can save you a lot of time and effort.

To use *fill down,* select a portion of a column. *Fill down* will take the top cell that has been selected and copy it into the lower cells. If the top cell in the selected range contains a number, such as 5, *fill down* will generate a column containing all 5s.

If the top cell of the selected range contains a formula, the *fill down* feature will automatically adjust the formula as you go from cell to cell.

File Edit	
Fill down	
Fill right	

	A	B
1	100	
2	A1+10	
3		
4		
5		
6		
7		
8		

File Edit	
Fill down	
Fill right	

	A	B
1	100	
2	A1+10	
3	A2+10	
4	A3+10	
5	A4+10	
6		
7		
8		

The selected column is highlighted.

The spreadsheet fills the column and adjusts the formula.

File Edit	
Fill down	
Fill right	

	A	B
1	100	
2	110	
3	120	
4	130	
5	140	
6		
7		
8		

These are the values that actually appear.

Fill right works in a similar manner, except it goes across, copying the leftmost cell of the selected range in a row.

File	Edit
Fill down
Fill right

	A	B	C	D	E
1	100				
2	A1+10				
3	A2+10				
4	A3+10				
5	A4+10				
6					
7					
8					

Row 1 is selected.

File	Edit
Fill down
Fill right

	A	B	C	D	E
1	100	100	100	100	100
2	A1+10				
3	A2+10				
4	A3+10				
5	A4+10				
6					
7					
8					

The 100 fills to the right.

If you select A1 to E1 and *fill right,* you will get all 100s.
If you select A2 to E2 and *fill right,* you will "copy" the formula A1 + 10 as shown.

File	Edit
Fill down
Fill right

	A	B	C	D	E
1	100	100	100	100	100
2	A1+10				
3					
4					
5					
6					
7					
8					

Row 2 is selected.

File	Edit
Fill down
Fill right

	A	B	C	D	E
1	100	100	100	100	100
2	A1+10	B1+10	C1+10	D1+10	E1+10
3	A2+10				
4	A3+10				
5	A4+10				
6					
7					
8					

The spreadsheet fills the row and adjusts the formula.

Check It Out
Use the last spreadsheet above.
10. "Select" B2 to B7 and fill down. What formula will be in B6? what number?
11. "Select" A3 to E3 and fill right. What formula will be in D3? what number?
12. "Select" C2 to C5 and fill down. What formula will be in C4? what number?

Spreadsheet Graphs

You can graph from a spreadsheet. As an example, let's use a spreadsheet to compare the **perimeter** of a square to the length of a side.

	A	B	C	D	E
		File Edit			

File   Edit

	A	B	C	D	E
1	side	perimeter			
2	1	4			
3	2	8			
4	3	12			
5	4	16			
6	5	20			
7	6	24			
8	7	28			
9	8	32			
10	9	36			
11	10	40			

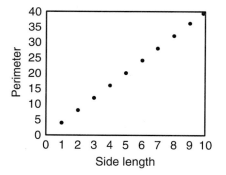

Most spreadsheets have a function that displays tables as graphs. See your spreadsheet reference for more information.

Check It Out

13. What cells gave the point $(2, 8)$?
14. What cells gave the point $(7, 28)$?
15. What point is shown by cells A11, B11?
16. What point is shown by cells A9, B9?

9.4 EXERCISES

For the spreadsheet shown to the right, what number appears in each of the following cells?

	A	B	C	D
1	1	1	200	1
2	2	3	500	6
3	3	5	800	15
4	4	7	1100	28
5				
6				
7				
8				

File Edit

1. A4
2. C3
3. B1

In which cell does each number appear?

4. 1100   5. 6   6. 5

7. If the formula behind cell B2 is B1 + 2, what formula is behind cell B3?
8. What formula might be behind cell A4?
9. If cells C5, C6, and C7 were filled in, what would the values of each of those cells be?
10. The formula behind cell D2 is A2 × B2. What formula might be behind cell D3?

Use the spreadsheet below to answer items 11–14.

File Edit

Fill down
Fill right

	A	B	C	D
1	5	10		
2	A1+6	B1×2		
3				
4				
5				
6				
7				
8				

11. If you select B2 to B6 and fill down, what formula will appear in B6?
12. Once you fill down as in item 11, what numbers will appear in B3 to B5?
13. If you select A2 to D2 and fill right, what will appear in D2?
14. If you select A1 to D1 and fill right, what will appear in C1?
15. If you select A1 to A7 and fill down, what will appear in A5?

What have you learned?

You can use the problems and list of words below to see what you have learned in this chapter. You can find out more about a particular problem or word by referring to the boldfaced topic number(for example, **9•2**).

Problem Set

Use your calculator for items 1–6. **9•1**
1. $27 + 11 \times 19$
2. 240% of 850

Round answers to the nearest tenth.
3. $70 - 18 \div (-10) + 43.1$
4. $14 + 58 \div 4.5 - 8.75$

Use the rectangle below for items 5 and 6.
5. Find the perimeter of rectangle $ABCD$.
6. Find the area of rectangle $ABCD$.

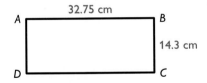

Use a scientific calculator for items 7–12. Round decimal answers to the nearest hundredth. **9•2**
7. 3.2^4
8. Find the reciprocal of 4.2.
9. Find the square of 4.2.
10. Find the square root of 42.25.
11. $(8 \times 10^6) \times (5 \times 10^7)$
12. $0.8 \times (55 \times 3.5)$

13. What is the measure of ∠SRV? **9•3**
14. What is the measure of ∠VRT? **9•3**
15. What is the measure of ∠TRS? **9•3**
16. Does R⃗T divide ∠VRS into two equal angles? **9•3**
17. What basic construction tools would you use to make an angle bisector? **9•3**

For items 18–20, refer to the spreadsheet. **9•4**
18. Name the cell holding 9.
19. A formula for cell B2 is B1 + 6. Name another formula for cell B2.
20. Cell D1 contains the number 240 and no formula. After using the command fill down on cells D1 through D8, what number will be in cell D8?

File Edit

Fill down
Fill right

	A	B	C	D
1	3	6	40	240
2	6	12	90	
3	9	18	140	
4				

WRITE THE DEFINITIONS FOR THE FOLLOWING WORDS.

hot words

angle **9•2**
arc **9•3**
cell **9•4**
circle **9•1**
column **9•4**
cube **9•2**
cube root **9•2**
decimal **9•1**
degree **9•2**

distance **9•3**
factorial **9•2**
formula **9•4**
horizontal **9•4**
negative number **9•1**
parentheses **9•2**
percent **9•1**
perimeter **9•4**
pi **9•1**
point **9•3**
power **9•2**

radius **9•1**
ray **9•3**
reciprocal **9•2**
root **9•2**
row **9•4**
spreadsheet **9•4**
square **9•2**
square root **9•1**
tangent **9•2**
vertex **9•3**
vertical **9•4**

WHAT HAVE YOU LEARNED?

Solutions

Index

SOLUTIONS

Chapter 1
Numbers and Computation

p. 70
1. 60,000 **2.** 6,000,000

3. $(2 \times 10,000) + (6 \times 1,000) + (7 \times 100) + (4 \times 10) + (8 \times 1)$ **4.** 666,718; 596,718; 66,718; 6,678 **5.** 62,574,860; 62,575,000; 63,000,000

6. 0 **7.** 18 **8.** 4,089 **9.** 0

10. 500 **11.** 1,400

12. $(4 + 7) \times 4 = 44$ **13.** $20 + (16 \div 4) + 5 = 29$

14. No **15.** No **16.** Yes **17.** No

18. 5×7 **19.** 5×23 **20.** $2^2 \times 5 \times 11$

21. 6 **22.** 15 **23.** 6

p. 71
24. 15 **25.** 200 **26.** 360 **27.** 12, 36, or 108

28. 8, 8 **29.** 14, -14 **30.** 11, 11 **31.** 20, -20

32. 7 **33.** -1 **34.** -12 **35.** 8 **36.** 0 **37.** 4

38. 35 **39.** -5 **40.** 6 **41.** 60 **42.** -24
43. -66

44. It will be a positive integer.

45. It will be a negative integer.

Place Value of Whole Numbers

1·1

1. 4,000 **2.** 40,000,000

3. Fifty million, three hundred twenty-six thousand, seven hundred **4.** Thirty-seven trillion, twenty billion, five hundred million

p. 73
5. $(9 \times 10,000) + (3 \times 1,000) + (4 \times 10) + (5 \times 1)$

6. $(6 \times 100,000) + (5 \times 100) + (8 \times 10) + (2 \times 1)$

p. 74   **7.** $<$ **8.** $>$ **9.** 6,520; 62,617; 67,302; 620,009
10. 27,400   **11.** 530,000   **12.** 4,000,000
13. 600,000

1•2 Properties

p. 76   **1.** Yes   **2.** No   **3.** No   **4.** Yes
p. 77   **5.** 26,307   **6.** 199   **7.** 0   **8.** 2.4
9. $(3 \times 2) + (3 \times 6)$ **10.** $6 \times (7 + 8)$
p. 78   **Number Palindromes**   One possible answer:
$21978 \times 4 = 87912$

1•3 Order of Operations

p. 80   **1.** 14   **2.** 81

1•4 Factors and Multiples

p. 82   **1.** 1, 2, 3, 6   **2.** 1, 2, 4, 8, 16
p. 83   **3.** 1, 2, 4   **4.** 1, 5
5. 4   **6.** 10
p. 84   **7.** Yes   **8.** No   **9.** Yes   **10.** Yes
p. 85   **11.** Yes   **12.** No   **13.** Yes   **14.** No
p. 86   **15.** $2^3 \times 5$ **16.** $2^2 \times 5^2$
p. 87   **17.** 2   **18.** 5   **19.** 3   **20.** 8
21. 24   **22.** 120   **23.** 72   **24.** 180
p. 88   **Darting Around**   One possible answer: $(3 \times 20 + 3$
$\times 20 + 2 \times 20) + (3 \times 20 + 3 \times 20 + 3 \times 20) +$
$(3 \times 20 + 3 \times 20 + 20) + (3 \times 20 + 3 \times 20 + 3 \times 20)$
$+ (50 + 50 + 3 \times 7) + (3 \times 20 + 3 \times 20 + 3 \times 20)$
$+ (2 \times 20)$

1•5 Integer Operations

p. 90   **1.** -4 **2.** $+300$
p. 91   **3.** 15, 15   **4.** 3, -3 **5.** 12, 12   **6.** 0, 0
7. -3 **8.** 0   **9.** -2 **10.** -4
p. 92   **11.** 8   **12.** -4 **13.** 4   **14.** -56

HOT SOLUTIONS

Chapter 2
Fractions, Decimals, and Percents

p. 98

1. 46.901 oz   2. 20 cups   3. 92%   4. 2.9 oz
5. C, $\frac{18}{60}$
6. $\frac{29}{30}$ 7. $1\frac{1}{7}$ 8. $3\frac{1}{2}$ 9. $5\frac{27}{35}$
10. B, $1\frac{1}{2}$
11. $\frac{3}{10}$ 12. $\frac{6}{35}$ 13. $1\frac{2}{7}$ 14. $1\frac{6}{7}$

p. 99

15. Thousandths   16. $2 + 0.002$
17. 300.303   18. 0.154; 1.054; 1.504; 1.540
19. 14.164   20. 1.44   21. 13.325   22. 34.8
23. 25%   24. 2   25. 92%
26. 99%   27. 40%
28. 7%   29. 83%
30. 0.17   31. 1.50 or 1.5
32. $\frac{27}{100}$ 33. $1\frac{1}{5}$

2·1 Fractions and Equivalent Fractions

p. 101   1. $\frac{3}{4}$ 2. $\frac{5}{9}$ 3. Answers will vary.
p. 103   4–6. Answers will vary.
p. 104   7. $=$ 8. $=$ 9. $\neq$
p. 105   10. 8; $\frac{3}{8}$; $\frac{2}{8}$ 11. 50; $\frac{35}{50}$, $\frac{21}{50}$
p. 106   12. $\frac{1}{7}$ 13. $\frac{1}{4}$ 14. $\frac{5}{6}$
p. 108   15. $3\frac{1}{6}$ 16. $1\frac{2}{3}$ 17. $1\frac{7}{9}$ 18. $7\frac{1}{8}$
19. $\frac{35}{4}$ 20. $\frac{61}{4}$ 21. $\frac{50}{3}$ 22. $\frac{59}{10}$

2·2 Comparing and Ordering Fractions

p. 111   1. $<$ 2. $>$ 3. $<$ 4. $>$
5. $<$ 6. $>$ 7. $<$
p. 112   8. $\frac{2}{3}, \frac{5}{7}, \frac{3}{4}, \frac{5}{6}$ 9. $\frac{2}{3}, \frac{3}{4}, \frac{7}{8}, \frac{9}{10}$ 10. $\frac{1}{8}, \frac{3}{8}, \frac{5}{12}, \frac{3}{4}, \frac{5}{6}$

2•3 Addition and Subtraction of Fractions

p. 114　**1.** $\frac{7}{5}$ or $1\frac{2}{5}$　**2.** $\frac{3}{2}$ or $1\frac{1}{2}$　**3.** $\frac{5}{9}$　**4.** $\frac{4}{13}$

p. 115　**5.** $1\frac{1}{8}$　**6.** $\frac{7}{30}$

p. 116　**7.** 6　**8.** $10\frac{1}{4}$　**9.** $5\frac{1}{4}$

p. 117　**10.** $7\frac{13}{18}$　**11.** $10\frac{1}{3}$

p. 118　**12.** $3\frac{5}{6}$　**13.** $2\frac{1}{2}$

2•4 Multiplication and Division of Fractions

p. 121　**1.** $\frac{3}{10}$　**2.** $1\frac{2}{3}$

　　　3. $\frac{3}{6} = \frac{1}{2}$　**4.** $\frac{4}{7}$　**5.** 4

　　　6. $\frac{5}{2}$　**7.** $\frac{1}{4}$　**8.** $\frac{3}{7}$

p. 122　**9.** $5\frac{1}{3}$　**10.** $3\frac{1}{2}$　**11.** $\frac{3}{4}$

2•5 Naming and Ordering Decimals

p. 125　**1.** 0.6　**2.** 0.44　**3.** 8.17　**4.** 3.4

5. Three hundred eighty-two thousandths　**6.** Two and one hundred fifty-four thousandths

7. Sixty-six thousandths

p. 127　**8.** Four ones; four and four hundred eleven thousandths　**9.** Two thousandths; thirty-two thousandths　**10.** Four thousandths; five and forty-six ten thousandths　**11.** One hundred thousandth; three hundred forty-one hundred thousandths

12. <　**13.** >

p. 128　**14.** 3.0186; 3.1608; 30.618　**15.** 9; 9.083; 9.084; 9.381; 93.8　**16.** 0.6212; 0.622; 0.6612; 0.662

17. 1.54　**18.** 36.39　**19.** 8.30 or 8.3　**20.** 8.68

HOT SOLUTIONS

2·6 Decimal Operations

p. 130 **1.** 39.85 **2.** 82.75 **3.** 13.11 **4.** 52.62781

p. 131 **Luxuries or Necessities?** About 923,000,000

p. 132 **5.** 14 **6.** 2 **7.** 14 **8.** 17

p. 133 **9.** 7.63 **10.** 0.5394

11. 1.952 **12.** 63.765

p. 134 **13.** 0.01517 **14.** 0.04805

p. 135 **15.** 270 **16.** 99 Answers may vary.

p. 136 **17.** 3.3 **18.** 5.2 **19.** 3.05 **20.** 0.073

21. 7.33 **22.** 101.56 **23.** 0.29

2·7 Meaning of Percent

p. 138 **1.** 36% **2.** 6% **3.** 50%

p. 139 **4.** About 50 **5.** About 25 **6.** About 75

7. About 10

p. 140 **8.** $.25 **9.** $17.00

Honesty Pays 20%

2·8 Using and Finding Percents

p. 142 **1.** 52 **2.** 425 **3.** 27 **4.** 315

p. 143 **5.** 29.64 **6.** 11.9 **7.** 80.52 **8.** 165

p. 144 **9.** $33\frac{1}{3}$% **10.** 450% **11.** 400%

12. 60%

p. 145 **13.** 104 **14.** 20 **15.** 25 **16.** 1,200

p. 147 **17.** 138% increase **18.** 88% decrease

19. 97% decrease **20.** 67% increase

p. 149 **21.** Discount: $36, Sale price: $54

22. Discount: $18, Sale price: $102

23. 22 **24.** 200 Answers may vary.

p. 150 **25.** $I = $195, A = 945 **26.** $I = $378, A = $3,978$

2·9 Fraction, Decimal, and Percent Relationships

1. 60% **2.** 30% **3.** 90% **4.** 208%

5. $\frac{17}{100}$ **6.** $\frac{1}{20}$ **7.** $\frac{9}{25}$ **8.** $\frac{16}{25}$

9. $\frac{129}{400}$ **10.** $\frac{191}{400}$ **11.** $\frac{985}{800}$

12. 27% **13.** 0.7% **14.** 1.8% **15.** 150%

Dollars, Pesos, Rupees, and Drachmas Australia, Japan, Kenya, and Mexico; $139.66

16. 0.49 **17.** 0.03 **18.** 1.8 **19.** 0.007

20. 0.3 **21.** 0.875 **22.** 0.$\overline{09}$

23. $\frac{39}{50}$ **24.** $\frac{27}{50}$ **25.** $\frac{6}{25}$

The Ups and Downs of Stocks 1%

Chapter 3
Powers and Roots

1. 3^5 **2.** a^3 **3.** 9^3 **4.** x^8

5. 9 **6.** 49 **7.** 16 **8.** 64

9. 27 **10.** 64 **11.** 216 **12.** 729

13. 10,000 **14.** 1,000,000 **15.** 10,000,000,000

16. 100,000,000

17. 5 **18.** 8 **19.** 10 **20.** 9

21. 5 and 6 **22.** 3 and 4 **23.** 8 and 9

24. 8 and 9

25. 6.928 **26.** 7.550 **27.** 9.434 **28.** 9.899

29. 3 **30.** 5 **31.** 6 **32.** 10

33. 3.6×10^7 **34.** 6×10^5

35. 8.09×10^{10} **36.** 5.4×10^2

37. 5,700,000 **38.** 1,998

39. 700,000,000 **40.** 734,000

HOT SOLUTIONS

3•1 Powers and Exponents

p. 166   **1.** 3^4 **2.** 7^4 **3.** a^6 **4.** z^5

p. 167   **5.** 16 **6.** 64 **7.** 9 **8.** 100

p. 168   **9.** 125 **10.** 1,000 **11.** 512 **12.** 216

p. 169   **13.** 100 **14.** 1,000,000

15. 100,000,000 **16.** 1,000

p. 170   **When Zeros Count** No; at the given rate, it would take almost 32 years to count to 1 billion and 10^{91} times as long to count to a googol.

3•2 Square and Cube Roots

p. 172   **1.** 3 **2.** 6 **3.** 9 **4.** 11

p. 173   **5.** Between 6 and 7 **6.** Between 4 and 5

7. Between 2 and 3 **8.** Between 9 and 10

p. 175   **9.** 1.732 **10.** 6.856 **11.** 9.274 **12.** 9.849

p. 176   **13.** 3 **14.** 7 **15.** 8 **16.** 10

3•3 Scientific Notation

p. 179   **1.** 5.3×10^4 **2.** 4×10^6 **3.** 7.08×10^{10}

4. 2.634×10^7

5. 67,000 **6.** 289,000,000

7. 170,300 **8.** 8,520,640,000,000

p. 180   **Bugs** 1.2×10^{18}—that is, one quintillion, two hundred quadrillion

Chapter 4
Data, Statistics, and Probability

p. 186   **1.** No   **2.** Biased
3. Stem-and-leaf plot   **4.** 38
5. You cannot tell from this graph.

6.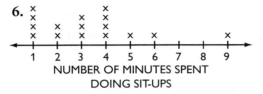

NUMBER OF MINUTES SPENT
DOING SIT-UPS

7.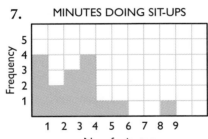

MINUTES DOING SIT-UPS

p. 187   **8.** 18 points   **9.** 87.4 points; 88 points   **10.** 5
11. 28   **12.** 5,040
13. $\frac{3}{5}$ **14.** 0   **15.** $\frac{3}{38}$

4·1 Collecting Data

p. 189   **1.** Adults registered to vote; 500   **2.** Fish in Sunshine Lake; 200

p. 190   **3.** No, it is limited to people who are in that store, so they may like it best.   **4.** Yes, each student has the same chance of being picked.   **5.** Answers will vary.

p. 191   **6.** It assumes you like mystery novels.
7. It does not assume you watch TV in the summer.
8. Do you read books?
9. 4   **10.** Factory   **11.** Science museum; more students chose that than any other possible field trip.

4•2 Displaying Data

p. 194

1. 7 and 8 letters

2.

Number of hours spent	0	1	2	3	4	5	6	7	8	9
Number of students	4	6	3	3	2	0	1	1	0	2

p. 195

3. 26   **4.** 25%   **5.** 50%

p. 197

6. About a quarter   **8.**

7. About half

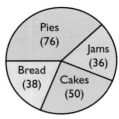

COUNTY FAIR ENTRIES

And the Winner Is...   Winning the award improved sales; weekly; bar graph

p. 198

9. 7   **10.** 4

11. NUMBER OF LETTERS IN EACH WORD

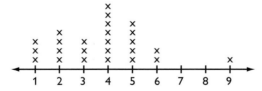

p. 199

12. 100   **13.** June

14. False; the chart records average numbers of ducks. To have an average of 100, there may have been more than 100 at some point.

p. 200

15. 12   **16.** 304   **17.** 22

p. 202

18. John Hancock Center

p. 202 (cont.)

19. NUMBER OF STUDENTS
AT RANDALL SCHOOL

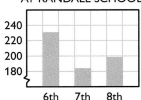

6th   7th   8th

20. 280   **21.** Answers will vary.

p. 204   **22.** 16   **23.**   NUMBER OF LETTERS
IN THE WORDS

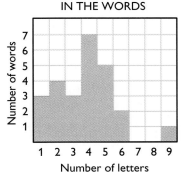

Graphic Impressions   Answers may vary. The graph
might be misinterpreted as being about the relative
sizes of the animals. The bar graph gives more
accurate information.

4.3 Analyzing Data

p. 207   **1.** Identifying Objects   **2.** No correlation
3. Distance Traveled

p. 208   **4.** Normal   **5.** Skewed left   **6.** Skewed right
7. Bimodal   **8.** Flat

4.4 Statistics

p. 211   **1.** 11   **2.** 78   **3.** 359.5   **4.** $52
p. 212   **5.** 11   **6.** 5.5   **7.** 48   **8.** 255 ft
p. 213   **9.** 27   **10.** 4.2   **11.** 11   **12.** $.75
p. 214   **13.** 910   **14.** 5.7   **15.** 66°   **16.** 15

4·5 Combinations and Permutations

p. 217 **1.** 9 **2.** 6 **3.** 8 **4.** 9

Monograms 17,576 monograms

p. 219 **5.** 24 **6.** 720 **7.** 24 **8.** 90 **9.** 40,320 **10.** 30

p. 221 **11.** 28 **12.** 220 **13.** 126 **14.** Six times as many permutations as combinations

p. 222 **Lottery Fever** Lightning (one in a million)

4·6 Probability

p. 225 **1.** $\frac{2}{5}$ **2.** $\frac{1}{10}$ **3.** Answers will vary.

p. 226 **4.** $\frac{1}{2}$ **5.** $\frac{1}{2}$ **6.** 0 **7.** $\frac{2}{11}$

p. 227 **8.** $\frac{3}{10}$; 0.3; 3:10; 30% **9.** $\frac{1}{4}$; 0.25; 1:4; 25%

p. 228 **10.** Red, white, red, yellow, red, red

11–12. Answers will vary.

p. 229 **13.**

	0	1	2	3	4	5	6	7	8	9
0	00	01	02	03	04	05	06	07	08	09
1	10	11	12	13	14	15	16	17	18	19
2	20	21	22	23	24	25	26	27	28	29
3	30	31	32	33	34	35	36	37	38	39
4	40	41	42	43	44	45	46	47	48	49
5	50	51	52	53	54	55	56	57	58	59
6	60	61	62	63	64	65	66	67	68	69
7	70	71	72	73	74	75	76	77	78	79
8	80	81	82	83	84	85	86	87	88	89
9	90	91	92	93	94	95	96	97	98	99

14. $\frac{1}{10}$

p. 230 **15.**

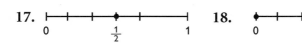

16.

17.

18.

p. 233   **19.** $\frac{1}{12}$; independent   **20.** $\frac{33}{95}$; dependent

         21. $\frac{16}{121}$ **22.** $\frac{6}{55}$

p. 234   **Flavor of the Week**   About 7 should prefer Banana Bonanza

Chapter 5 Logic

p. 240   **1.** True   **2.** False   **3.** False   **4.** False   **5.** True

      6. True   **7.** False   **8.** True   **9.** True

     10. If it is after 7 P.M., then the streetlights are on.

     11. If an angle is acute, then it has a measure greater than 0° but less than 90°.

     12. If $a^2 = 16$, then $a = 4$.   **13.** If I arrive at school on time, then the bus is running on schedule.

     14. The street is not closed for repairs.   **15.** These two lines are perpendicular.

p. 241   **16.** If $x + 7 \neq 12$, then $x \neq 5$.   **17.** If you do not do all your homework, then you will not receive a good grade.

     18. If you do not become lazy, then you do not watch too much television.   **19.** If the outcome was not heads or tails, then you did not flip this coin.

     20. Labor Day   **21.** 12

     22. $\{a\}, \{b\}, \{c\}, \{a, b\}, \{a, c\}, \{b, c\}, \{a, b, c\}, \varnothing$

     23. $\{11, 12, 13, 14, 15\}$ **24.** $\{11, 12, 13, 14\}$

     25. $\varnothing$ **26.** $\{5, 15\}$

     27. $\{1, 3, 5, 7, 9\}$ **28.** $\{5, 9, 13, 17\}$

     29. $\{1, 3, 5, 7, 9, 13, 17\}$ **30.** $\{5, 9\}$

5·1 If/Then Statements

p. 243   **1.** If two segments have the same length, then they are congruent.   **2.** If an integer ends in 0, then the integer is a multiple of 10.

Continued

p. 243
(cont.)

3. If you study algebra, then you are in the seventh grade. **4.** If you got a sum of 7, then you added 3 and 4.

p. 244

5. 5 is not greater than 4. **6.** We will not go to the beach this summer. **7.** If an integer does not end with 0, then it is not a multiple of 10. **8.** If you do not live in San Diego, then you do not live in the state of California.

p. 245

9. If a number is not divisible by 3, then it is not divisible by 6. **10.** If two lines intersect, then they are not parallel.

p. 246

Rapunzel, Rapunzel, Let Down Your Hair If there are more than 150,000 people in your town, there must be two people with the same number of hairs on their heads.

5•2 Counterexamples

p. 248

1. True; False; counterexample: a 16° angle

2. True; True

5•3 Sets

p. 250 **1.** False **2.** True **3.** $\{4\}, \{8\}, \{4, 8\}, \varnothing$ **4.** $\{m\}, \varnothing$

p. 251 **5.** $\{3, 6, 8, 10\}$ **6.** $\{x, y\}$ **7.** $\{14\}$ **8.** $\varnothing$

p. 252 **9.** $\{1, 2, 3, 4\}$ **10.** $\{2, 4\}$ **11.** $\{2, 3, 4, 5, 6, 7\}$

12. $\{3, 4\}$ **13.** $\{4\}$

Chapter 6 Algebra

p. 258

1. $2x - 5 = x + 3$ **2.** $6(n + 2) = 2n - 4$

3. $4(x + 6)$ **4.** $3(3n - 4)$ **5.** $6a + 2b$ **6.** $5n - 7$

7. 10 mi **8.** $x = 6$ **9.** $y = -18$ **10.** $x = 5$

11. $y = 35$ **12.** $n = 5$ **13.** $y = -4$

14. $n = 8$ **15.** $x = -2$ **16.** 15 girls **17.** 7 cm

p. 259

18.

```
←——+——+——+——+——○——+——+——+——→
  -5  -4  -3  -2  -1   0   1   2
                x < -1
```

19.

```
←——+——+——+——+——●——+——+——→
  -1   0   1   2   3   4   5
                x ≥ 3
```

20–23.

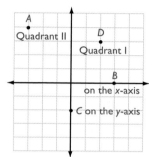

24. $\dfrac{-5}{3}$

25.

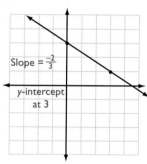

Slope = $\dfrac{-2}{3}$

y-intercept at 3

26.

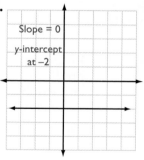

Slope = 0

y-intercept at -2

27.

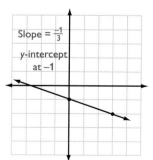

Slope = $\dfrac{-1}{3}$

y-intercept at -1

28.

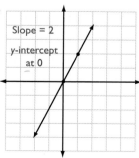

Slope = 2

y-intercept at 0

continued

HOT SOLUTIONS

p. 259 (cont.) **29.** $y = x - 3$ **30.** $y = \frac{3}{2}x + 1$

6·1 Writing Expressions and Equations

p. 260 **1.** 2 **2.** 1 **3.** 3 **4.** 2

p. 261 **5.** $3 + x$ **6.** $n + 9$ **7.** $y + 5$ **8.** $n + 4$

p. 262 **9.** $10 - x$ **10.** $n - 7$ **11.** $y - 5$ **12.** $n - 8$

p. 263 **13.** $6x$ **14.** $4n$ **15.** $0.75y$ **16.** $10n$

p. 264 **17.** $\frac{x}{3}$ **18.** $\frac{12}{n}$ **19.** $\frac{30}{y}$ **20.** $\frac{a}{7}$

p. 265 **21.** $5n - 8$ **22.** $\frac{2}{x} - 4$ **23.** $2(n - 10)$

Orphaned Whale Rescued $2{,}378 + 25d = 9{,}000$

p. 266 **24.** $3x - 6 = 4x + 2$ **25.** $\frac{x}{4} + 5 = x - 10$

26. $3n - 2 = 2(n + 3)$

6·2 Simplifying Expressions

p. 268 **1.** No **2.** Yes **3.** No **4.** Yes

5. $3 + 2x$ **6.** $5n$ **7.** $3y + 6$ **8.** $6 \cdot 5$

p. 269 **Predicting Life Expectancy** This equation would predict a life expectancy of 400 years by the year 3000 and would also indicate that the life expectancy was zero years in the year 1750.

p. 270 **9.** $3 + (7 + 10)$ **10.** $4 \cdot (2 \cdot 7)$ **11.** $4x + (3y + 5)$

12. $(4 \cdot 5)n$

13. $6(100 - 2) = 588$ **14.** $3(100 + 5) = 315$

15. $4(200 - 1) = 796$ **16.** $5(200 + 10 + 4) = 1{,}070$

p. 271 **17.** $10x + 4$ **18.** $18n - 12$ **19.** $-6y + 4$

20. $8x - 10$

p. 272 **21.** $7(x + 3)$ **22.** $3(4n - 3)$ **23.** $10(c + 3)$

24. $5(2a - 5)$

p. 274 **25.** $13x$ **26.** $4y$ **27.** $10n$ **28.** $-4a$
29. $5y + 6z$ **30.** $7x - 15$ **31.** $4a + 2$ **32.** $7n - 4$

6·3 Evaluating Expressions and Formulas

p. 276 **1.** 14 **2.** 2 **3.** 15 **4.** 14
p. 277 **5.** 26 cm **6.** 50 m **7.** 14 in. **8.** 24 ft
p. 278 **9.** 30 mi **10.** 2,100 km **11.** 330 mi **12.** 10 ft
Maglev $1\frac{1}{4}$ hr, $2\frac{1}{4}$ hr, $3\frac{3}{4}$ hr

6·4 Solving Linear Equations

p. 280 **1.** -4 **2.** x **3.** 35 **4.** $-10y$
p. 281 **5.** True, false, false **6.** False, true, false
7. False, true, false **8.** False, false, true
9. Yes **10.** No **11.** No **12.** Yes
p. 282 **13.** $x + 3 = 15$ **14.** $x - 3 = 9$ **15.** $3x = 36$
16. $\frac{x}{3} = 4$
p. 283 **17.** $x = 7$ **18.** $n = 13$ **19.** $y = -6$ **20.** $a = 6$
p. 285 **21.** $x = 5$ **22.** $y = 20$ **23.** $n = -3$ **24.** $a = 30$
p. 286 **25.** $x = 5$ **26.** $y = 50$ **27.** $n = -4$ **28.** $a = -6$
p. 287 **29.** $n = 4$ **30.** $x = -2$
p. 289 **31.** $n = 5$ **32.** $x = -1$
33. $w = \frac{A}{l}$ **34.** $y = \frac{-x + 6}{3}$
p. 290 **Three Astronauts and a Cat** 79 fish

6•5 Ratio and Proportion

p. 292   **1.** $\frac{4}{8} = \frac{1}{2}$ **2.** $\frac{8}{12} = \frac{2}{3}$ **3.** $\frac{12}{4} = \frac{3}{1} = 3$

p. 293   **4.** Yes   **5.** No

p. 294   **6.** 4.5 gal   **7.** $350

      Prime Time   About 43,500,000

6•6 Inequalities

p. 297   **1.**

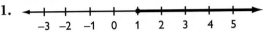

      2.

      3.

      4.

p. 298   **5.** $x > -5$ **6.** $n \le -4$ **7.** $y < 3$ **8.** $x \ge 1$

6•7 Graphing on the Coordinate Plane

p. 300   **1.** x-axis   **2.** Quadrant III   **3.** Quadrant I

      4. y-axis

p. 301   **5.** $(4, -2)$ **6.** $(-1, 3)$ **7.** $(2, 0)$ **8.** $(0, -1)$

p. 302   **9–12.**

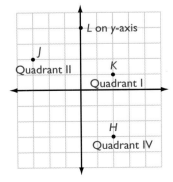

p. 304   13–16.

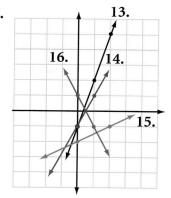

13.
16.   14.
15.

p. 305   17–20.

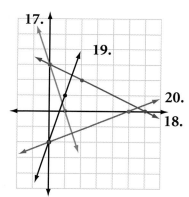

17.
19.
20.
18.

p. 306   21–24.

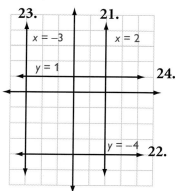

23.   21.
x = −3   x = 2
y = 1   24.

y = −4   22.

6•8

Slope and Intercept

p. 309   **1.** $\frac{3}{2}$ **2.** $\frac{-3}{4}$

p. 310   **3.** −1 **4.** $\frac{3}{2}$ **5.** $\frac{-1}{2}$ **6.** 4

p. 311   **7.** 0 **8.** No slope **9.** No slope **10.** 0

p. 312   **11.** 0 **12.** 4

p. 313

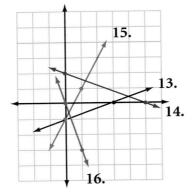

p. 314   **17.** Slope $= -3$, y-intercept at 2   **18.** Slope $= \frac{1}{4}$, y-intercept at -2 **19.** Slope $= \frac{-2}{3}$, y-intercept at 0   **20.** Slope $= 6$, y-intercept at -5

p. 315   **21–24.** graphs: see equations below.

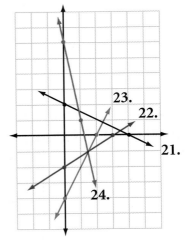

21. $y = \frac{-1}{2}x + 2$ **22.** $y = \frac{2}{3}x - 2$

23. $y = 2x - 4$ **24.** $y = -5x + 6$

p. 316

25.

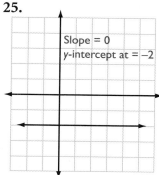

26.

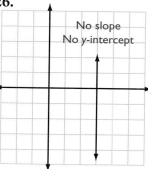

27.

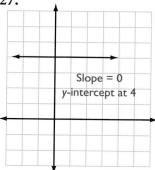

28.

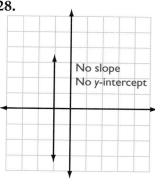

p. 317   **29.** $y = 2x - 4$ **30.** $y = \frac{-2}{3}x + 2$ **31.** $y = 2x + 1$
32. $y = \frac{-1}{2}x - 1$

p. 318   **33.** $y = -x + 5$ **34.** $y = \frac{2}{9}x + \frac{2}{3}$

Chapter 7   Geometry

p. 324   **1.** $62°$ **2.** $\overrightarrow{AB}, \overrightarrow{AD}$ **3.** $360°$ **4.** $28°$

5. $108°$ **6.** H, X, O   **7.** 14 cm   **8.** 38 m
9. 60 ft^2 **10.** 56 in.^{2} **11.** 70 cm^2 **12.** 1,256 m^2
13. 125 in.^{3} **14.** 80 cm^3 **15.** 480 cm^3

p. 325   **16.** $120°$ **17.** 16π m^2
18. 5 in.   **19.** 12 in.   **20.** 0.75

7•1 Naming and Classifying Angles and Triangles

p. 326   **1.** $\overleftrightarrow{KL}$, $\overrightarrow{LK}$ **2.** K

p. 328   **3.** N **4.** $\angle KNL$ or $\angle LNK$, $\angle LNM$ or $\angle MNL$, $\angle KNM$ or $\angle MNK$

p. 329   **5.** 60°   **6.** 85°   **7.** 145°

p. 330   **8.** 45°; acute angle   **9.** 180°; straight angle
10. 135°; obtuse angle

p. 332   **11.** $m\angle L = 30°$ **12.** $m\angle C = 45°$
13. C and D   **14.** A

7•2 Naming and Classifying Polygons and Polyhedrons

p. 335   **1.** Possible answers: *ABCD*; *ADCB*; *BCDA*; *BADC*; *CDAB*; *CBAD*; *DABC*; *DCBA*   **2.** 360°   **3.** 65°

p. 336   **4.** No; yes; no; yes; no   **5.** Yes; it has four sides that are the same length and opposite sides are parallel.

p. 338   **6.** Yes, hexagon   **7.** Yes, quadrilateral   **8.** No

p. 340   **9.** 900°   **10.** 120°

p. 341   **11.** Pentagonal prism   **12.** Square pyramid

7•3 Symmetry and Transformations

p. 345   **1.**

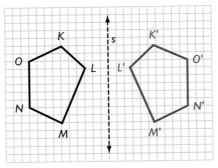

Flip It, Slide It, Turn It Messages   Can math be fun?

p. 346   **2.** No   **3.** Yes, one   **4.** Yes, two   **5.** Yes, two

p. 347   **6.** 270°   **7.** 180°

p. 348   **8.** Yes; 7 units right and 2 units up   **9.** No

        10. Yes; 7 units left and 4 units up

7·4 Perimeter

p. 351   **1.** $16\frac{1}{2}$ in.   **2.** 56 m   **3.** 4 ft   **4.** 21 m

p. 352   **5.** 30 ft   **6.** 36 in.

        The Pentagon   924 ft

p. 353   **7.** 36 m   **8.** 30 cm

7·5 Area

p. 356   **1.** About 30 m^2

p. 358   **2.** 324 $in.^2$ **3.** 169 m^2

        4. 48 cm^2 **5.** 11 ft

p. 359   **6.** 75 cm^2 **7.** 120 $in.^2$

p. 360   **8.** 25 m^2 **9.** 17 cm^2

7·6 Surface Area

p. 363   **1.** 250 $in.^2$ **2.** 22 m^2

p. 364   **3.** 900 m^2 **4.** B   **5.** 408.2 $in.^2$

7·7 Volume

p. 366   **1.** 5 m^3 **2.** 3 cm^3

p. 367   **3.** 420 ft^3 **4.** 48 $in.^3$

p. 368   **5.** 62,800 ft^3 **6.** 1,570 cm^3

p. 370   **7.** 160 cm^3 **8.** 392.5 $in.^3$ or 392.7 $in.^3$

        Good Night, T. Rex   About 1,175,000 mi^3

7•8 Circles

1. 13 m   **2.** 0.5 cm   **3.** 32 in.   **4.** 5 ft   **5.** 12 m
6. 28π ft   **7.** 21π cm
8. 45.8 m   **9.** 113 cm   **10.** 6.5 in.
11. $\angle WXY$ or $\angle YXW$ **12.** 60°   **13.** 70°
Now, That's a Pizza!   About 2,500 ft^2
14. 169π cm^2 **15.** 346 ft^2

7•9 Pythagorean Theorem

1. 9, 16, 25   **2.** Area A + Area B = Area C
3. 8 m   **4.** 20 cm

7•10 Tangent Ratio

1. 5.5 cm   **2.** 58°

Chapter 8   Measurement

1. Possibilities include gram, kilogram; ounce,
pound, ton.   **2.** Possibilities include millimeter,
centimeter, meter, kilometer; inch, foot, yard, mile.
3. Customary
4. 4,500 m   **5.** 8,400   **6.** 144 in.   **7.** 0.25 mi
8. 360 in.   **9.** 10 yd   **10.** 5,184 in.^2 **11.** 4 yd^2
12. 2,380 mm^2 **13.** 63 ft^2 **14.** 1,296 in.^2
15. 1,900,000 cm^3 **16.** 8 yd^3
17. 5,000,000 mm^3 **18.** 0.005 m^3
19. About 305 in.^2
20. 3.75 lb   **21.** 12 packets   **22.** 30,240 min
23. 31,536,000 sec in a regular year and
31,622,400 sec in a leap year
24. 6 in.   **25.** $\frac{4}{9}$

8·1 Systems of Measurement

p. 393   **1.** Metric   **2.** Customary

p. 394   **3.** The area is about 18.5 cm^2. The actual answer lies between 18.06 cm^2 and 18.84 cm^2, so the answer can't be given as 18.49 cm^2.   **4.** No; the answer is given to more decimal points than the distance traveled. The actual answer is about 44.6 mi/hr.

8·2 Length and Distance

p. 396   **1–2.** Answers will vary.

p. 397   **3.** 2.2 km   **4.** 5 ft

p. 398   **5.** 78.7 cm   **6.** 70 yd   **7.** A   **8.** B

8·3 Area, Volume, and Capacity

p. 400   **1.** 63 ft^2 **2.** 40,000 cm^2

p. 401   **3.** 0.125 m^3 **4.** 10 yd^3

p. 402   **5.** The quart

8·4 Mass and Weight

p. 404   **1.** 4,200 lb   **2.** 0.64 g   **3.** 88.2 lb

p. 405   **Poor SID**   No, the mass is always the same.

8·5 Time

p. 406   **1.** 527,040 min   **2.** Friday

p. 407   **The World's Largest Reptile**   5 times as long

8·6 Size and Scale

p. 408   **1.** Q and U are similar. R and T are similar.

p. 409   **2.** $\frac{5}{2}$ **3.** $\frac{3}{4}$

p. 410   **4.** $\frac{25}{16}$ **5.** 100 ft^2

Chapter 9   Tools

1. 151   **2.** 2,550
3. 21.8   **4.** −77.6
5. 58.5 cm   **6.** 179.375 cm^2
7. 262.14   **8.** 0.19   **9.** 67.24   **10.** 2.86
11. 1.4 × 10^9 **12.** 84
13. 120°   **14.** 60°   **15.** 60°   **16.** Yes
17. Compass
18. B2   **19.** A2 × B2   **20.** 2,000

9·1 Four-Function Calculator

1. 26.7   **2.** 7.9   **3.** −5   **4.** −40
5. 160   **6.** 355   **7.** 8,021   **8.** 217
The Mystery of Memory   Answers may vary.
Immediate memory is used for entering the numbers;
short-term memory is used for remembering the
running total; and long-term memory is used for
remembering the rules and arithmetic facts.
9. 30   **10.** 580   **11.** 22   **12.** 87.5

9·2 Scientific Calculator

1. 40,320   **2.** 14,641
3. 0.125   **4.** −3,637
5. 1.6   **6.** 0.93
Magic Numbers   12,345 × 8 + 5 = 98,765

9·3 Geometry Tools

1. 7.6 cm or 3 in.   **2.** 7 cm or $2\frac{3}{4}$ in.   **3.** 2 in. or
5.1 cm   **4.** 4.1 cm or $1\frac{5}{8}$ in.
5. 28°   **6.** 110°

p. 431   **7–10.** Measure radius to check. Remember that radius is the distance from the center of the circle to the outside edge.

p. 432   **11.** Each angle is 22°.

12.

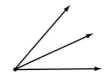

13. 45°   **14.** Measure the two angles created by the bisector and make sure they are each half the measurement of the original angle.

9•4 Spreadsheets

p. 436   **1.** 1   **2.** 9   **3.** 16
          4. False   **5.** False

p. 437   **6.** B3 × C3   **7.** B4 × C4   **8.** $40   **9.** $36

p. 439   **10.** B5 + 10; 150   **11.** D2 + 10; 120   **12.** C3 + 10; 130

p. 440   **13.** A3, B3   **14.** A8, B8
          15. (10, 40)   **16.** (8, 32)

INDEX